Food Science and Technology

A Resource for Year 11 General

Yvette Alexander
Julie Luscombe
Elaine McNally

Food Science and Technology
A Resource for Year 11 General

IMPACT PUBLISHING
PO Box 200
Cottesloe
Western Australia 6911

Telephone: (08) 9286 4209
Fax: (08) 9286 4207

Email: info@impactpublishing.com.au
Internet: www.impactpublishing.com.au

DISCLAIMER

This text has been developed to assist both teachers and students in meeting the required Outcomes for the Food Science and Technology Year 11 General course. Although the information presented in this resource is accurate to the best of their knowledge, the authors cannot guarantee that every statement is without flaw of any kind.

Therefore, the authors disclaim all liability for any errors, or for any loss or other consequences resulting from any individual relying on, or acting upon, any information provided in this resource.

Although every effort has been made to trace the sources of information used in the preparation of this book, if we have not acknowledged adequately any person to their satisfaction, we apologise. Please inform us by emailing info@impactpubishing.com.au and the matter will be rectified for any further editions or published works on this course in the future.

COVER DESIGN
Leanne Quince, GraphicsAbove

PRINTING
Smart-EL

ISBN 978 1 921965 89 0

First Edition 2014

Printed in China by 1010 Printing International Limited

Acknowledgements

We would like to thank our publishers Regina Gaujers and Gail Warrilow for their patience, creativity, advice and assistance to facilitate the publication of this text.

To our fellow Home Economics teachers at Churchlands Senior High School we thank you for your continued professionalism and inspiration. Your support has been very much appreciated. Thank you also the students at Churchlands for giving us permission to use their photos.

Lastly, to our families, thank you for your encouragement, patience and assistance during the rewriting of this text and with assisting in correcting any issues that we had.

Contents

INTRODUCTION

In Australia we are being exposed to a continually evolving food culture. Our melting pot of ethnicities have merged to create an exciting fusion of cuisines, and exposure to reality media contests have ignited interest in the home chef creating more varied and exciting menus. This surely can only bode well for the food and hospitality industry as food becomes more of an explorative lifestyle choice than simply a mundane daily need. Supermarket shelves now carry thousands of ingredient selections and Australia's primary producers are providing us with choice of quality ingredients. Our food industry is becoming an exciting and valuable aspect of our Australian culture.

The importance of food in the many aspects of our daily life makes this subject Food Science and Technology developed by the Curriculum Council of Western Australia a very worthwhile subject where there are many opportunities to learn valuable life skills.

This text has been designed to provide knowledge about many aspects of food and the food industry. Our use of and attitude to food is often changing as are our needs through life. Many options exist when making food choices and balancing the use of raw and processed food products continually challenges our food related decisions.

The advancement of technology has seen continual change in food products, food developments and food systems and services. Manufacturers continually seek market share by creating new products and enticing consumers using carefully planned strategies that may or may not be to your benefit. You are encouraged to explore and question the viability of these technological advancements.

In this text we have provided you with information about many aspects of food and the food industry, instructions for food preparation skills such as the basic cookery methods and mise en place suitable for use in industry and the home. The food preparation skills and information will be valuable for those students wishing to study the hospitality course.

We hope this text gives you an insight into the importance of food and the food industry.

Enjoy,
The Authors

Unlocking the chapters

The chapters have been written in sections that are taken from the course content. They are mostly titled in the relevant category, but some extra chapters have been inserted in the 'Processing Food' category to include relevant information deemed important for completion of the course.

Each chapter begins with a list of key concepts which describe the content that is covered in the chapter.

The chapter content then focuses on the breakdown of this subject matter with a combination of information and activities, to ensure you cover all aspects. As you work through the chapters you will notice different aspects of the book.

These include:

- Activities
- Definitions
- 'Did you know'
- Extension activities
- General information.

Activity:

Discussions, debates and interesting activities are suggested.

Definitions: Important terms and definitions will be given.

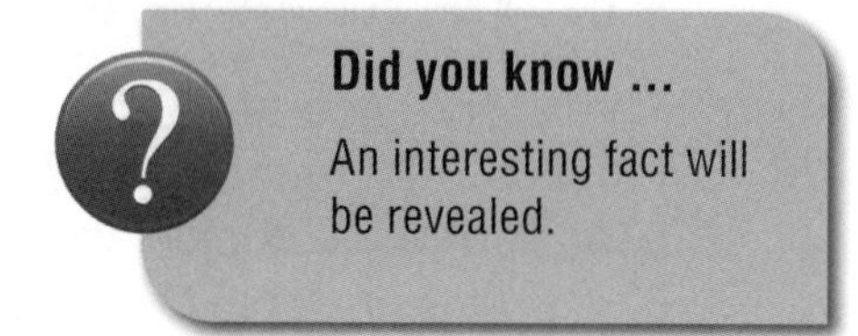

At the end of the chapter, Extension Activities are suggested as larger projects that may take one or more lessons to complete.

Enjoy this book and we hope you make the most of your cooking experiences throughout the year.

CHAPTER 1
Food as a Commodity

Key Concepts

- **Classification: raw foods and processed foods**
- **Classification: plant foods**
  - **cereals**
  - **vegetables**
  - **fruit**
  - **legumes, pulses seeds, nuts**
  - **oils**
- **Classification: animal foods**
  - **eggs**
  - **meat**
  - **poultry**
  - **seafood**
  - **milk**
- **Classification: staple foods**
- **Environmental and economic factors when purchasing food**

INTRODUCTION

In this chapter we investigate the various types of natural foods, and the environmental and economic factors that contribute to the availability of these foods.

Students will look at many methods of classifying foods such as by the raw food groups, natural versus processed, animal versus plant, and staple foods.

The raw materials (food) are called **commodities** as they can be bought and sold for profit. **Food processing** is where raw food is turned into products that may be eaten.

Over the years global migration has influenced the food we eat and has contributed to the introduction of many different foods into the Australian diet. There is also an increasing trend in the use of native Australian foods in our food habits.

Equipment, storage, production, manufacturing and distribution techniques together with market place practices have changed what we eat. The affluence of the economy also affects the food that is available.

CLASSIFICATION: RAW FOODS/ PROCESSED FOODS

A. Raw foods

Foods that have undergone minimal processing and contain no preservatives or artificial colouring are considered to be natural foods. Fresh fruit and vegetables would be considered raw foods. However, once they are manipulated, for example canned, the food is classified as being processed. On most occasions it is best to choose the raw food rather than the processed variety. The quality and nutritional value is generally preferable in the raw food. There are many unanswered questions about the use of additives in processed food and there is a definite trend in recent times of turning back to raw foods. However in some circumstances, such as the food processing of milk, it makes it safer for users to consume.

B. Processed foods

Foods that have been transformed from their original state using various methods and techniques are considered processed foods. The properties of the food are changed to preserve it, improve the quality or make it more useful or stable.

Did you know ...

The more ingredients listed on a label generally means the more the food has been processed ?

There are two types of food processing:

1. **Primary processing.** This involves the simpler methods of processing such as chopping and mincing to cut food up (eg. meat), the cleaning of fish and milling of flour.
2. **Secondary processing.** This involves changing primary products into other edible food products. An example is milk. Primary processing occurs when the milk is pasteurised then secondary processing occurs when the milk is used for making ice cream, butter or cheese.

Activity 1.1: Processing

1. What are some other examples of primary food processing?
2. What are some other examples of secondary food processing?

Food processing can change the colour, aroma, flavour and texture of food. The changes to the food are both physical and chemical and some changes are reversible. For example, chocolate melts when heated then solidify again when cooled. Most processing changes are, however, irreversible, for example cake mixes and batters.

Thousands of food items are processed each year.

The aims of processing are to:

- make the food safe by eliminating micro-organisms
- improve the quality (colour, texture, flavour)
- make food into forms that are easier to use and more convenient.

Activity 1.2: Food processing

1. What other advantages of processing food can you think of?
2. What are some of the disadvantages of processed food? Can you think of any?

Advantages of processed food are:

- toxins are removed
- seasonal foods are available all year round
- food is of a consistent standard
- perishable food maybe transported over long distances
- food is safe to eat.

Food scientists are continually striving to develop products that are of a high quality, and adapting food processing techniques. Consumers are demanding foods that have been processed without the addition of preservatives, artificial colouring and flavours, and appear fresh with a similar resemblance to the natural food.

The problem is the nutrient value of food is often altered by processing.

Some food processing techniques include:

- canning
- dehydration/drying
- pasteurisation
- freezing
- blanching
- milling
- freeze drying.

Activity 1.3: Methods of food processing

1. What other methods of food processing can you think of?

Let's look at the process involved with some of these techniques.

1. **Freezing:** This is a common method of food processing which slows food decay, and by the conversion of water into ice the food becomes unavailable for bacterial growth. This also slows down chemical reactions.

   As freezing only slows down enzyme reactions often foods that need to be frozen are blanched or chemicals are added to stop the enzyme activity before freezing.

   Foods that are frozen will keep for several months providing they are kept at -18°C.

2. **Canning:** This form of food processing preserves the food by sealing it in an airtight container. It prevents micro-organisms from entering and damaging the food inside.
3. **Drying:** This is the oldest method of food processing known from when sun and wind were used in ancient times. Water is removed from the food which prevents the growth of micro organisms and decay. Many different foods may be dehydrated. Examples of these are meat, fruit, onions and chillies.

Activity 1.4: Food processing techniques

1. Define blanching.
2. When purchasing canned food what advice would you give a consumer to ensure the food is not damaged inside the can?
3. What is fruit leather? How is it produced?
4. Explain freeze drying.
5. Much has been debated about the quality of commercially processed food. It often contains too much salt, sugar, fat and additives.

   Complete the following table outlining wise choices when purchasing processed food.

Processed food to purchase	Processed food to avoid
	• Some canned foods that contain lots of salt • White breads pastas and other products made from refined flour • High kilojoule snack food • High fat convenience food • Packaged cakes and biscuits • Processed meat • High sugar, high salt, low fat breakfast cereals

6. What processed foods are best nutritionally to purchase?

Activity 1.5: Processed tomatoes

Years ago fresh tomatoes were only available in summer. Now they are available all year around in many different varieties. With technological advancement tomatoes are now available in many forms on the supermarket shelves. Go to your local supermarket and investigate all the various methods of purchasing tomatoes. Consider the cost per kilogram for each method.

What are the uses, disadvantages and advantages of the various forms of processed tomatoes and what is the impact on the environment?

CLASSIFICATION: PLANT FOODS

The plant tissues that are consumed are fruits, vegetables, cereals, seeds, roots, cereals (wheat, rice, maize, oats), legumes, seeds and nuts.

A. Cereals

Grains are the edibles seeds of grasses.

Grains produced in Western Australia include cereals (wheat, rice, triticale, rye, corn, oats and barley), pulses (lupins, chickpeas, mung beans, lentils) and oilseeds (canola, sunflowers, soybeans).

These cereals are used in the production of flour, breakfast cereals, bread, pasta and noodles, cakes, biscuits and pastries, glucose, malt, beer and animal feed. Many of the breads that are made today have a combination of many grains and cereals, for example multigrain bread. They are considered to be a non-perishable food due to their low moisture content. When stored correctly they have a long shelf life.

Did you know ...

Buckwheat is not a cereal but is related to rhubarb and sorrel. It is often used in porridge and the seeds may be ground into flour for use in pancakes.

Quinoa is commonly considered a cereal but it is actually a relative of leafy green vegetables like spinach. It is a protein rich seed with a fluffy, creamy appearance, crunchy texture and slightly nutty flavour.

Nutritional value of cereals

- **Carbohydrates:** approximately 58-72% of cereal composition is made up of carbohydrate, present mainly in the endosperm.
- **Fats:** small amounts are present, especially in the germ.
- **Protein:** 7-13% of the cereal is protein and it is found in the endosperm.
- **Vitamins:** cereals are a good source of the Vitamin B group including thiamine, riboflavin, folate and niacin and are found in the germ and bran layer.
- **Minerals:** iron, phosphorous and calcium are present in some cereals. These are found in the germ and bran layer.
- **Soluble fibre:** this is found in oats and barley and has been shown to lower blood cholesterol.

Activity 1.6: Whole grains

Why are whole grain cereals considered to be more nutritious?

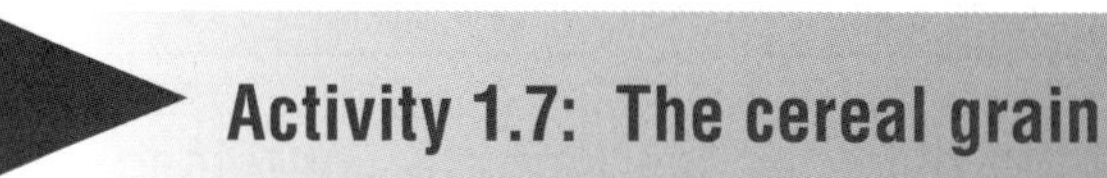

Activity 1.7: The cereal grain

Draw a picture of a cereal grain and label the following parts: bran, endosperm and germ. Use the Internet or library reference books to find a suitable diagram.

Serving sizes

One serving size that is based on the recommended daily intake for adults can be described as:

- 2 slices bread
- 1 small potato
- $1\frac{1}{3}$ cups cereal
- 4 tablespoons muesli
- 1 muffin
- 1 cup cooked porridge
- 1 crumpet
- 1 cup cooked pasta, rice or noodles.

Types of grains

1. Wheat

This is the most common cereal worldwide and is used in the making of breakfast cereals and flour. Gluten is found in flour to varying degrees and this is what gives flour its elastic properties and strength in the final food product.

Bread flour has a higher gluten content than regular flour hence it is stronger. This improves the structure of the bread. It is for this reason that it is recommended that bread flour is used for choux pastry.

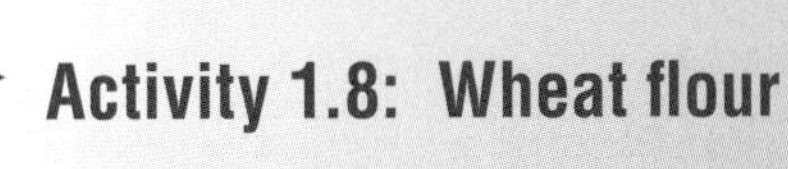

Activity 1.8: Wheat flour

1. **Experiment:** Take a piece of muslin 30 cm square. Into this piece of muslin, place 1 cup of flour and twist the top of the muslin. Rinse the flour under running water until the water is no longer cloudy. What remains is gluten. Repeat using bread flour and gluten flour. What is the difference between these three types of flour?
2. Investigate the differences between hard and soft flour. What are the uses for each of these types of flour?
3. In the processing of wheat grain various methods of milling produce a range of wheat products. Complete the following chart.

Wheat Flour Products	Description
Plain flour	
Self-raising flour	
Wholemeal flour	
Wholegrain flour	
Wheaten flour/cornflour	
Wheat germ/bran	
Couscous	
Gluten flour	
Semolina	
Kibbled wheat	
Bulgur	

4. Design a couscous salad suitable to be served at the school canteen. It must be appealing to teenagers and be of reasonable cost. It should contain ingredients that allow it to be prepared one day and served the next. What changes would you make if it was to be served at a primary school canteen?

2. Rice

Rice is an important cereal grain eaten all around the world. Rice grains vary in shape and size making them suitable for different cooking methods. It is sold as brown rice (whole grain) or as white rice (endosperm only).

Rice products

Rice is found in breakfast cereals as rice flakes or popped rice and in rice cakes, rice noodles and rice flour or ground rice. Rice flour is used to give a gritty texture to products such as shortbread.

Activity 1.9: Types of rice

Complete the following table.

Types of rice	Description	Uses
Arborio		
Basmati		
Brown rice		
Calrose		
Jasmine		
Quick cook rice		
White rice		
Wild rice		
Koshihikara rice		

Cooking rice

Rice maybe cooked by the following methods – boiling, microwave, electric rice cooker and the absorption method. When cooking rice it is important to maintain the grain's structure and avoid over-swelling or breaking.

3. Corn

Corn is also known as maize and is harvested as a cob. The kernels are stripped off the cob and eaten as a vegetable or the kernels are milled to produce corn products.

Products from processing include cornflakes used as a breakfast cereal, popcorn, cornflour, corn syrup and cornmeal.

Activity 1.10: Shortbread comparisons

1. Produce and compare two shortbread recipes, one using rice flour and one without rice flour.

Butter Shortbread

125 gms butter
¼ cup icing sugar
¾ cup flour

1. Preheat oven to 180 °C.
2. Have butter at room temperature.
3. Place butter, icing sugar and flour into a food processor and process the mixture until it combines together.
4. Roll out and cut into shapes.
5. Bake until pale golden.
6. Stand for a few minutes before cooling on a wire rack.

Shortbread

125 gms butter
1 tablespoon ground rice or rice flour
1 cup plain flour
2 tablespoons icing sugar

1. Preheat oven to 160 °C.
2. Chop chilled butter and place in a food processer with the remaining ingredients. Process until all the ingredients come together.
3. Roll out the mixture and cut into shapes.
4. Bake until pale golden.
5. Stand for a few minutes on the tray then cool on a wire rack.

2. Complete a sensory evaluation of the two different shortbreads. What difference does the rice flour make? Which do you prefer?

Activity 1.11: Cooking rice

1. Prepare a cup of rice by each of these methods and complete the chart following.

Feature	Boiling	Microwave	Electric rice cooker	Absorption
Appearance				
Flavour				
Texture				
Advantages				
Disadvantages				

2. Which method did you prefer and why?
3. Why is it recommended that rice is washed before cooking in some recipes?

Activity 1.12: Corn

1. Name three products, and their country of origin, that may be made from cornmeal or ground corn.
2. What are Hominy Grits?

4. Rye

Rye is milled to produce a dark flour usually used in bread. As it contains no gluten the bread is often dense and heavy.

Dark rye flour is used for pumpernickel bread and the lighter rye flour is used for bread and biscuits. It may also be combined with wheat flour in the development of food products.

Rye flour is also used for crisp breads.

5. Oats

Most oats are used in breakfast cereal, cakes and biscuits.

Rolled oats are flattened long grains which are used in breakfast cereals like porridge, and in slices and biscuits. Anzac Biscuits are a common biscuit in which rolled oats is one of the main ingredients.

Oats that are milled to form small particles for use in breads, flour products and breakfast cereals are called oatmeal.

Activity 1.13: Oats

1. How do quick cook oats and instant oats differ from rolled oats?
2. Prepare rolled oats and quick cook oats as a breakfast cereal. Do a sensory evaluation of both these products. Devise your own evaluation chart. Which type of oat cereal do you prefer? Why?
3. Using rolled oats as a base, create a hot winter cereal breakfast dish suitable for your family. Write out your recipe.

6. Triticale

This is a combination of wheat and rye. It has a much higher protein content and better balance of essential amino acids than either wheat or rye. Triticale has a nut-like taste and is either flaked or milled into flour. It has only recently been developed into a viable crop and may be purchased at health food shops or is found in some breakfast cereals.

7. Barley

This is sold as pearl barley which is used in soups and casseroles and, more recently, to replace rice in risotto. It was a very popular cereal many years ago especially when used in soups. Recently there has been resurgence in its use in dishes such as risotto and soups.

Activity 1.14: Barley

1. Try the following recipe using pearl barley in a risotto.

   Chicken and Pearl Barley Risotto
   (Serves 2)

1 Tbsp olive oil	4 cups chicken stock
200 gms chicken breast	2 Tbsp chopped parsley
1 medium onion chopped	2 Tbsp chopped basil
1 stick celery sliced	2 cups English spinach
½ red capsicum chopped	½ cup grated strong cheddar cheese
1½ cups pearl barley rinsed	

   1. Heat oil in a heavy based saucepan. Add chicken breast and cook, then slice into pieces. Set aside.
   2. Add onion, celery and capsicum to the pan and cook until soft. Add pearl barley and chicken stock. Bring to boil and then lower heat to simmer the risotto with the lid on the pan.
   3. Cook until the liquid is absorbed or the pearl barley is soft. Add extra chicken stock if needed.
   4. Remove from heat and stir through chicken, spinach, parsley and basil.
   5. Return to a low heat and gently reheat risotto. Stir through cheese, season to taste and serve immediately.

2. What variations could you make to this recipe?
3. How does the taste and texture of pearl barley compare to rice when used in a risotto?

Selection and storage of cereals

When purchasing cereal products check for any insect infestation (webs in the packet) and the use-by date.

- Packages should be tightly sealed and free from tears.
- Once opened the cereal products should be stored in an airtight container in a cool dark place.
- Avoid contact with moisture and dampness when storing.
- Always empty and wash the storage containers before refilling with a new packet of cereal.
- Avoid contact with moisture when storing.

B. Vegetables

Vegetables are the edible parts of plants. Vegetables may be classified by their colour, how they are used or, most commonly, by the part of the plant that is used; for example, the celery stalk is used so it is a stem vegetable.

Our early vegetable intake was influenced by our European settlers. It was a limited selection due to what seeds or plants were being brought out on ships from the British Isles. Our present vegetable varieties have increased considerably due to our multicultural society, improved methods of the importation of vegetables and improved agricultural techniques within Australia.

Vegetables may be served as an accompaniment to meat, poultry and fish dishes, as a basis for a main course or as snacks. They add flavour, colour, texture and nutritional value to meals.

Vegetables may be baked, boiled, braised, fried, grilled, roasted, sautéd and steamed. They may be purchased fresh, frozen, dehydrated and canned.

See Table 1.1 on the following page for classification of vegetables and how they are selected, prepared and stored.

C. Fruit

Fruit is the edible part of a plant that contains a seed or the matured ovary of a flower. They are high in vitamins and minerals, are low in fat, contain small amounts of protein and vary in the amounts of carbohydrate present. Fruit has a high water content and is low in kilojoules.

TABLE 1.1: Classification of vegetables

Edible part of plant	Vegetable	Selection	Preparation	Storage
Bulbs	Onions (red, white, brown), spring onions, leek, garlic, fennel	Firm even size, skin feels dry. Leeks and spring onions should be bright green in colour and the leaves crisp.	Remove outside skin and base. Remove any discoloured leaves or dry ends of leeks and spring onions. Wash if necessary.	Onions and garlic are best stored in a cool dark place. Leeks and spring onions should be stored in the crisper of a fridge.
Roots	Carrots, parsnip, beetroot, turnip, swede, radish	Firm, smooth, crisp	Wash, peel, top and tail.	Store in the crisper section of the fridge.
Tubers	Potatoes, sweet potato (kumara), jerusalem artichokes	Firm, smooth, even size. No greening on the potato skin and free from shoots	Wash and use peeled or unpeeled.	Store in a cool, dark place.
Stems	Celery, asparagus, bok choy	Crisp, free from cracks	Wash asparagus; remove the hard end of the stalk. Celery – wash and remove the coarse strings.	Store in the crisper section of the fridge.
Leaves	Brussel sprouts, cabbage, spinach, silver beet, lettuce	Crisp, bright green colour	Wash and remove outer leaves of cabbage and brussel sprouts.	Store in the crisper section of the fridge.
Flowers	Cauliflower, broccoli, zucchini flowers, globe artichoke, marigolds, nasturtiums	Compact heads, cauliflower white heads, broccoli dark green with a purplish tinge	Wash and remove thick tough stalk of cauliflower and broccoli.	Wrap in plastic and store in the crisper section of the fridge.
Fungi	Mushrooms, truffles	Firm, white to cream colour, dry surface	Wipe clean with a damp paper towel. Brush with a pastry brush.	Store in a paper bag in the lowest part of the fridge.
Seeds	Fresh – peas, broad beans, sweet corn, snow peas Dried – lentils, pulses	Crisp, green, full pods Free from impurities	Remove seeds from pod.	Store in the crisper section of the fridge.
Sprouts	Shoots from seeds, eg. mung beans	Crisp, green	Wash.	Store in the crisper section of fridge.
Fruits	Avocado, cucumber, eggplant, capsicum, squash, marrow, zucchini, tomato, pumpkin, chillies, beans	Crisp, smooth skin	Wash and peel where necessary.	Store in the crisper section of fridge.

Activity 1.15: Classification of fruit

Next to each of the classifications give examples of fruit found in this category, selection points and storage points.

Classification	Examples	Selection	Storage
Berry			
Citrus			
Dried			
Exotic			
Melons			
Pomes			
Stone fruit			
Tropical			
Vines			

Nutritional value of fruit and vegetables

The fibre and nutrient content of all fruit and vegetables is higher if they are unpeeled, for example apples.

- **Vitamins:** a very good source especially when the fruit and vegetables are fresh. Citrus fruit is high in Vitamin C and Vitamin A is found in orange and yellow vegetables.
- **Minerals:** a very good source.
- **Carbohydrates:** present in the form of starches and sugars in varying amounts
.

Serving sizes

It is recommended that a person eats two pieces of fruit and five to seven different vegetables per day.

TWO pieces of fruit and FIVE-SEVEN different vegetables per day!

1 serving size: VEGETABLES

½ cup cooked vegetables
1 corn cob
1 cup salad vegetables
1 medium potato

1 serving size: FRUIT

1 medium apple, pear, banana, orange
2 plums
1 cup fruit salad or stewed fruit
1 tablespoons sultanas

Selection and storage of fruit and vegetables

1. To ensure freshness purchase fruit and vegetables from a supplier that has a high turnover.
2. Fruit and vegetables should not be bruised, discoloured or damaged.
3. Fruit and vegetables that are in season have a better flavour and should be cheaper in price.
4. Fruit and vegetables should be crisp and colourful.
5. Only purchase amounts that are needed because of their perishable state.
6. Discard any fruit and vegetables that become mouldy during storage.
7. Most fruit and vegetables should be stored in the refrigerator. Exceptions to this are potatoes, onions and garlic which should be stored not in a plastic bag but rather in a cardboard box in a cool dark place.

Preparation and cooking

1. Fruit and vegetables should be washed thoroughly before use to remove contaminants such as dirt, bacteria and chemical residue.
2. Fruit and vegetables should be cut into same sized pieces for even cooking.
3. Vegetables and fruit maybe cooked by boiling, braising, steaming, poaching, stir-frying, baking, roasting, grilling, microwaving and stewing.
4. Fruit should be prepared just before use to prevent discolouration, especially with apples, pears and bananas. Fruit and vegetables should be cooked with care to prevent the colour and texture from being spoilt.

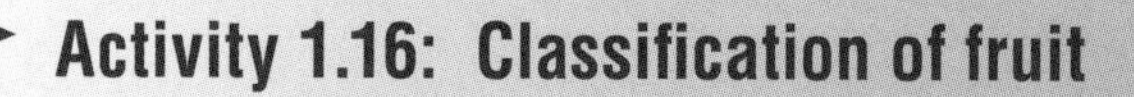

Activity 1.16: Classification of fruit

1. For each of the following cooking methods write down five examples of fruit and/or vegetables that are suited to that cooking method.

   - Boiling
   - Braising
   - Microwaving
   - Stewing
   - Steaming
   - Roasting
   - Baking
   - Poaching
   - Stir-frying
   - Grilling

2. From a nutritional aspect why is it recommended that some fruit and vegetables are eaten fresh each day?

3. Design and prepare a Super Vegie Pattie suitable for a vegetarian. What points will you consider in the planning? How will you serve your patty attractively? What herbs and flavourings will you add? (Remember to use them in moderation.) Write out a detailed recipe. What do you have to consider when preparing and cooking the vegetables?

4. You have been asked to put away the family's fruit and vegetable shopping. Firstly, you will discard any fruit and vegetables that have become mouldy and damaged. For each of the following fruit and vegetables, indicate how you would store them to maintain maximum freshness.

   - Carrots
   - Lettuce
   - Bananas
   - Potatoes
   - Grapes
   - Apples
   - Brown onions
   - Whole pumpkin
   - Tomatoes
   - Oranges
   - Spring onions
   - Sweet potato
   - Mushrooms
   - Fresh coriander

5. Fruit and vegetables maybe purchased dried, canned or frozen. What are the advantages of using the fruit and vegetables in these forms? What are some of the disadvantages?

6. How can you stop the browning of fruits such as pears, apples and bananas?

D. Legumes, pulses, seeds and nuts

Legumes are often known as peas, beans and lentils and may be purchased fresh, frozen, canned or dried. Some examples of these are mung beans, chickpeas, kidney beans and soy beans. Legumes are an economical source of protein and may be added to a meal in a variety of ways, such as in salads, patties or as a separate vegetable serve.

Dried legumes should be soaked overnight then cooked for a long period of time.

Seeds are the embryo and food supply of young plants whereas nuts are dried tree fruits and are contained within a hard shell.

Pulses are the edible seeds of the legume family. Some examples are butter beans, broad beans and cannellini beans. They may be purchased dried or in cans. Can you think of some other seeds that we eat?

Seeds and nuts are a valuable food source as they are high in protein.

Nutritional value

- **Protein:** legumes,nuts, seeds are a rich source but they do not contain all the essential amino acids.
- **Fat:** low source.
- **Vitamins:** rich source.
- **Minerals:** rich source.
- **Carbohydrate:** good source.
- **Fibre:** high source.

Selection and storage

1. Legumes should be clean, dry and free flowing.
2. Packaging should be free of damage and there should be no sign of insect infestation.
3. Dried legumes should be stored in an airtight container in a cool, dark, dry place.
4. Once cooked, legumes should be stored in a covered container in the fridge.

Activity 1.17: Legumes

1. What products are made from soya beans?
2. Why are legumes good for diabetics?

E. Oils

Oils are of plant origin and are liquid at room temperature. Some examples of these are olive oil, canola oil and sunflower oil. Some oils, such as the cold press of olive oil or similar minimum processed oils, can be considered a raw food.

Oils are high in polyunsaturated fats.

CLASSIFICATION: ANIMAL FOODS

The animal classification includes both red and white meats, organs, eggs and milk.

A. Eggs

Eggs are a versatile food and are nutrient dense. In Australia, the eggs consumed are mainly from hens with smaller quantities from ducks, quails, geese and bantams. Quail eggs are used mainly for garnishing and pickling. Duck eggs have a stronger flavour. Eggs may be prepared in a variety of simple methods or used in more complex recipes.

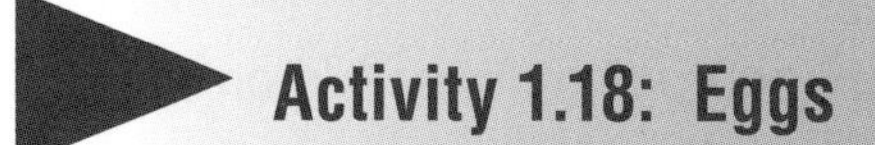

Activity 1.18: Eggs

What is the difference between caged eggs, free range eggs and barn eggs?

Nutritional content of eggs

- **Protein:** eggs are rich in high quality protein.
- **Vitamins:** vitamins A, D and E are found in the egg yolk and most of the B group vitamins are in both the yolk and white.
- **Minerals:** iron is the most important mineral found in eggs. Other minerals found in eggs are zinc, iodine, phosphorous, potassium and smaller quantities of others.
- **Fat:** some fat is found in the egg yolk.

Egg structure and composition

An egg's weight is usually 58 gms but may vary from 35 gms to 77 gms.

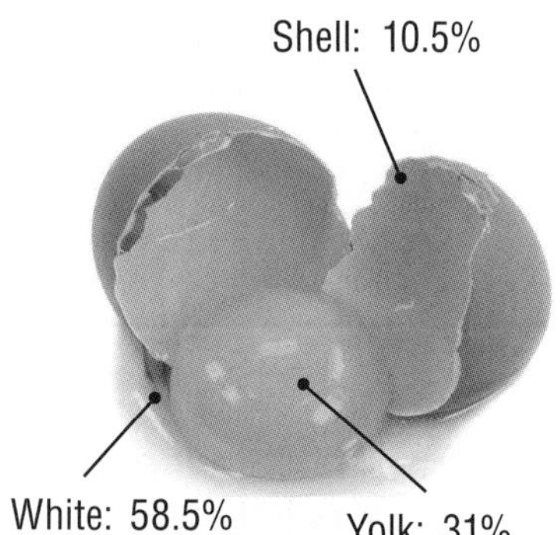

1. Shell

- Is mainly calcium carbonate.
- May be brown or white, which has no influence on the nutritional value of the egg.
- Is covered in approximately 8,000 to 10,000 tiny pores which allow moisture and gas exchange.
- Prevents bacteria entry into the egg.

2. Shell membranes

- One membrane sticks to the shell and the other surrounds the white.
- Protect the egg against bacteria.
- are composed of thin layers of protein.

3. Germinal disc

- Provides the entry for fertilisation of the egg.

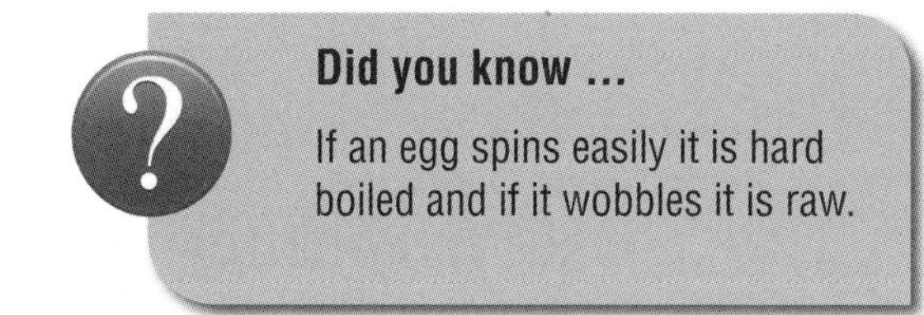

4. White (albumen)

- has two layers : thick and thin.
- Consists of mostly water, high protein and some minerals.

5. Chalaza

- Are spiral bands that anchor the yolk to the centre of the thick albumen.

6. Yolk membrane

- Surrounds the yolk.

7. Yolk

- Has a colour that ranges from pale yellow to dark orange depending on the hen's diet.
- Is a major source of vitamins, minerals, protein and essential fatty acids.

8. Air cell

- Forms at the wide end of the egg.

Selection of eggs

- Shells should be free from dirt and cracks.
- Check use-by dates on the carton.

Storage of eggs

- Eggs are best stored in their original container in a refrigerator. The cartons prevent water loss and other foods being absorbed into the eggs.
- Eggs will keep for approximately four weeks, refrigerated.
- Keep eggs stored pointed end down so that the yolk stays centred and prevents damage to the air sack at the blunted end.

Freshness test

- A fresh raw egg in its shell will sink in a basin of water.
- The smaller the air cell the fresher the egg.
- The chalaza is prominent in a fresh egg.
- When a fresh egg is broken, the thick albumen stands up firmly around the yolk.
- The thick albumen is easily distinguishable from the thin albumen.

Uses in cooking

Eggs are a versatile food and provide a variety of functions in food preparation. Eggs may be a meal in themselves when poached, scrambled, fried, boiled or made into an omelette.

Other uses of eggs are outlined below.

- **Binding:** eggs hold ingredients together, for example when making meatballs.
- **Clarifying agent:** egg white is an effective clarifying agent used in the preparation of jellies and consommés.
- **Coating:** eggs may be used as a coating for fried foods and they also help coatings such as bread crumbs and couscous stick to food. The egg prevents the fat soaking into the food, for example in crumbed fish.
- **Emulsifier:** eggs stabilise oil and vinegar in mixtures such as mayonnaise.
- **Foam:** air is incorporated into mixtures such as cakes, meringues and soufflés through the beating and aeration of eggs.
- **Glaze:** baked goods such as pastries take on a golden appearance when brushed with a beaten egg. A lightly beaten egg white brushed on a baked pastry case and then baked for a minute prevents the case from becoming soggy when filled.
- **Raising agent:** eggs assist to incorporate air into mixtures, for example cakes and soufflés.
- **Setting:** when eggs coagulate with heat they set fillings such as custards and quiches.
- **Stabilise:** egg whites coat sugar crystals or air bubbles preventing the formation of large sugar crystals and/or the formation of large ice crystals.
- **Thickening:** Eggs give a smooth and creamy texture to sauces and soups as well as enrich the flavour, for example béarnaise sauce.

Did you know ...

The protein in eggs provides the stretch and increase in volume. When heated the egg white sets to a mass and the yolk becomes firmer. The setting process is called coagulation and is due to denaturation of the proteins. Eggs should be cooked over a gentle heat otherwise they will toughen and become hard. Remember: **Denaturation** is the irreversible change in protein's structure. **Coagulation** is the clotting of protein.

Activity 1.19: Egg challenges

1. You have a special guest staying overnight. In the morning you decide to prepare a breakfast for this person. You have available to use 2 eggs, ½ tomato, 6 mushrooms and 2 slices of bread. There is also access to milk, butter, herbs and spices in the pantry. With a time limit of 20 minutes plan, prepare and serve this meal attractively. Describe what you made and the process you used. Evaluate your end result discussing how you were successful and how you could improve if you were to repeat this exercise.

2. How do you prevent stirred egg custards and baked egg custards from curdling?

B. Meat

Meat is the skeletal muscle and internal organs of birds and animals that is eaten by humans. The quality of the meat depends on the animal's age, breed, diet and handling in the slaughtering process. Meat may be purchased fresh, salted, smoked, as sausages or in a vast range of smallgoods, for example salami and pepperoni. It is very high in protein and over the years farmers have modified farming techniques to reduce the fat content. Meat may be cattle (beef, veal), sheep (lamb, mutton, hogget), pig (pork, ham, bacon), poultry (chicken, duck, turkey, goose) and game (rabbit, kangaroo, emu, venison, crocodile, quail, pheasant.) Many cooking methods are suitable for cooking meat; the method used depends on the degree of toughness of the meat.

Activity 1.20: Meat

1. Cattle provide beef and veal – can you explain the difference? Discuss the appearance of the meat.

2. Sheep provide lamb, mutton and hogget – can you explain the difference? Discuss the appearance of the meat.

Structure of meat

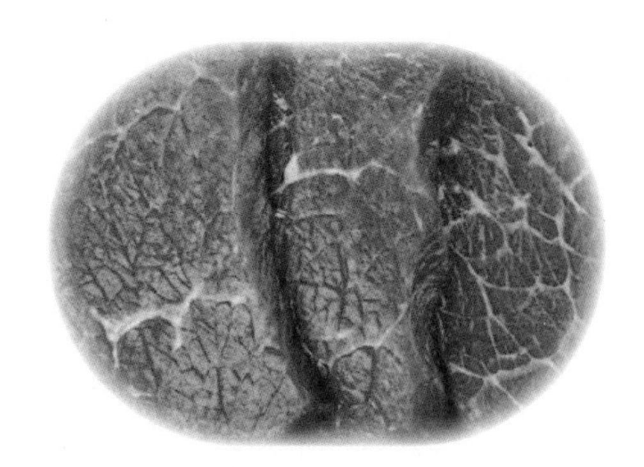

Meat consists of fatty tissue, muscle fibres, connective tissue and bone. The muscle fibres provide meat with its characteristic texture; the larger the fibres the coarser the texture. The muscle fibres are surrounded by a layer of connective tissue which is made up of collagen fibres. Fatty tissue is found around and within the muscle fibres and in the connective tissue. It also forms around some organs to protect them, for example kidneys.

The tenderness of meat depends on:

- the age of the animal and which part it comes from. The older the animal the tougher the meat
- the more activity the part of the body has done the tougher the meat, for example, the muscles of the leg and neck are tougher.

The fibres in tender meat are short and fine whereas in tough meat the fibres are thicker with more connective tissue.

Did you know ...

Meat is an expensive part of a family's food costs therefore it is important that the correct cooking method is chosen for the cut of meat. A darker colour of fat usually indicates the meat is from an older animal.

Activity 1.21: Cooking meat

1. Complete the following chart, indicating the tough and tender cuts.

Type	Tough cuts of meat	Tender cuts of meat
Beef		
Lamb		
Pork		

2. Give examples of suitable cooking methods for tender cuts of meat.
3. Give examples of suitable cooking methods for tough cuts of meat.
4. Explain marbling.
5. Explain how wet cooking methods tenderise tough cuts of meat.

Nutritional value of meat

- **Protein:** meat is an excellent source of high quality protein. About 20% of the muscle tissue is made up of essential protein. One hundred to 125 gms of meat in combination with another protein rich food is adequate to supply the body's daily requirement of protein (40 to 50 gms). Excess protein intake is used for energy or stored in the body as fat.
- **Fats:** the fat found in meat is in the form of saturated fat therefore it is advisable to trim meat of any visible fat when preparing it. The fat is a source of fat soluble vitamins and provides the body with a source of energy, so the fat marbled throughout the muscle tissue provides this requirement easily.
- **Vitamins:** meat is a good source of the B group vitamins, especially B12 which assists the nervous system.
- **Minerals:** meat is an excellent source of iron. It also provides zinc which is important for a healthy immune system. Potassium, magnesium and phosphorous are also found in meat.
- **Carbohydrates:** very little is found in meat.
- **Water:** meat is made up of approximately 75% water.

Serving sizes

One to two servings are recommended per day. A serving size is equivalent to 65 to 100 gms of meat. Examples of serving sizes are:

- ½ cup mince
- 2 small chops
- 2 slices roast meat.

Effect of cooking on meat

The application of heat causes both physical and chemical changes to meat. These are outlined below.

1. Most bacteria are destroyed therefore rendering meat safe to eat.
2. The meat changes colour due to the alteration of colour pigments.
3. Flavour is developed making meat more palatable and appetising.
4. Water is evaporated causing the meat to shrink.
5. Fat melts which increases the flavour and helps in the browning of meat.
6. Slow moist cooking methods soften the collagen converting it to gelatine, making it easier to chew and digest.

Marinades and meat mallets may be used to tenderise meats. Meat mallets will physically breakdown the muscle fibres and connective tissue making them shorter and hence easier to digest. As well as tenderising the meat a marinade will also add flavour.

Storage of meat

- Fresh meat should be wrapped in plastic film and stored in the coldest part of the refrigerator.
- Meat may be wrapped in plastic film or sealed in a plastic bag and frozen for longer periods of time. Always freeze meat in meal size lots, labelled with the name of the cut of meat and dated. Thaw in the fridge and do not refreeze.
- Cooked meat using the wet cooking methods may also be frozen. When heated the meat should be reheated to a high temperature to prevent food poisoning.

Further, the place of purchase should be clean.

- Ensure meat is displayed and stored at the correct temperature.
- Avoid meats with excessive amounts of fat.
- The surface of meat should be moist, but not feel slimy, and have a pleasant fresh odour.

Activity 1.22: Meat enhancing and storing

1. Explain the reactions that occur when meats are tenderised by marinades. Research some different types of marinades and explain the role of each of the ingredients.
2. Explain what is barding and larding.
3. What is freezer burn and how may it be prevented?

C. Poultry

Poultry has gained in popularity over the years due to its flavour, versatility in preparation, having less saturated fat than red meat and its reasonable price. Poultry includes chicken, duck, turkey and goose.

Nutritional value of poultry

- **Protein:** poultry has the same amount of protein as red meat.
- **Minerals:** it is a good source of zinc, phosphorous, magnesium and a moderate source of iron.
- **Fat:** it contains more polyunsaturated fat and less saturated fat than red meat. The skin of the bird is high in fat and it is advisable to remove it to make the meal healthier.
- **Vitamins:** they are present in small amounts.

Structure of poultry

Chicken has a finer texture, less connective tissue and less fat than red meat.

Poultry may be cooked in pieces or as a whole bird. Darker meat indicates it is from the active muscles of the birds and it is older. Chicken may also be purchased as wings, drumsticks (legs), breast, tenderloins and thighs. The pieces may have the bone in or taken out.

Cooking of poultry

Chicken has tender flesh therefore may be prepared by poaching, roasting, grilling, frying, stewing, casseroling and barbecuing. Chicken maybe purchased as the whole bird, or in portions such as breasts, legs, chicken fillets, thighs, and wings. The healthier option is to buy the product as skinless or remove the skin before preparing.

Cooking increases the tenderness of the meat, improves the flavour and kills any harmful micro-organisms.

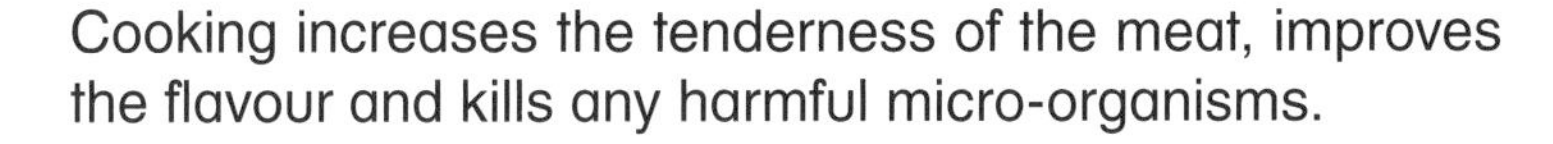

Selection of poultry

- Frozen chicken should have no frozen juices around the outside of the bird.
- The skin on fresh chicken should be plump with no dry patches or tears.
- The breast should be plump.
- The flesh should be firm, smell fresh and have no film on it.
- It should have a fresh odour.

Activity 1.23: Chickens

1. If you purchased a chicken maryland what parts of the chicken would you be purchasing?
2. When purchasing chickens they are sold as No. 9, No. 12 and so on. Explain what this means.
3. Think of a meal suitable for a family meal using the whole chicken. The person preparing the meal only has 1 hour to make the recipe. If you had to prepare this what part of the chicken would you test to ensure it is cooked? Why?

Storage of poultry

- Chicken pieces should be wrapped and stored in the fridge.
- Chicken pieces should be wrapped and stored in meal size lots and dated before freezing.
- Chicken should be thawed in a fridge.

Please Note: Chicken is highly susceptible to food poisoning therefore it must be cooked until there are no pink juices flowing from the chicken. Check the chicken is completely thawed before cooking. Always store chicken, both cooked and uncooked, in a fridge. Wash and pat dry the carcass of a whole chicken before cooking.

Activity 1.24: Cooking chickens

For each cut, name the most preferable methods of cooking.

- Breast
- Thigh
- Wings
- Drumsticks

D. Seafood

Australia has an abundance of high quality seafood. Until recently, most of our seafood supplies have been from their natural environment; however, fish farming or aquaculture is now becoming increasingly important as a seafood provider. Some fish are available all year round while other varieties are seasonal. Some varieties are available only in certain waters.

Fish may be purchased fresh, frozen, smoked, dried and canned.

Nutritional value

- **Protein:** fish contains complete protein which means the protein contains all the essential amino acids.
- **Fat:** fat ranges from 0.7 to 18 %. The fat found in fish is considered to be desirable as it contains omega 3 fatty acids.
- **Minerals:** seafood may contain calcium if the bones are eaten and iodine and fluorine from salt water fish.
- **Carbohydrates:** carbohydrates are very low in fish.
- **Vitamins:** vitamin A is found in oily fish.

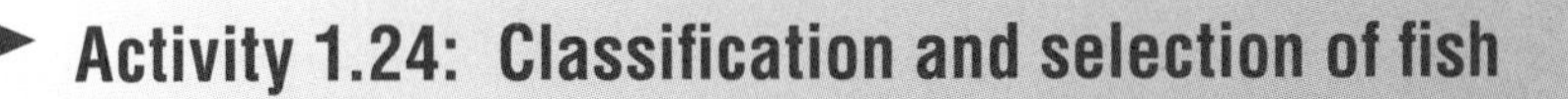

Activity 1.24: Classification and selection of fish

1. Fish maybe classified as:
   - Oily fish
   - White fish
   - Shellfish

   NOTE: Seafood may also be classified as freshwater (eg. marron, bream) and saltwater(herring, prawns).

   Give examples of each.
2. What parts of a fish which you would check for freshness?
3. Fish may also be purchased canned, dried, smoked and frozen. When purchasing frozen fish what are some of the points that should be considered?

Activity 1.25: Purchasing fish

1. When purchasing fish what is the difference between a fish cutlet and a fish fillet?
2. If you saw sashimi on the menu what would it mean?

Selection of seafood

Fresh fish may be purchased as whole, fillets or cutlets. Consider the following criteria when selecting each type of seafood.

Whole fish

1. The eyes of fish should be bright and bulging, not cloudy and sunken.
2. Flesh should be firm and springy.
3. The gills should be bright pink/red in colour.
4. Fish should have a pleasant fresh sea smell.
5. Tails and fins should be moist and pliable.

Fillets and cutlets

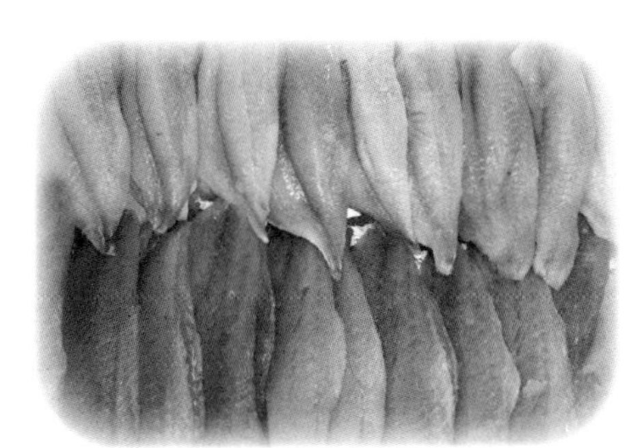

1. Flesh should be bright, firm and lustrous.
2. There should be no discolouration or bruising.
3. Fish should have a pleasant fresh sea smell.
4. The edges of the fish should not be dried out.

Shellfish

1. Flesh and shells should be brightly coloured and lustrous.
2. Shells, heads, tentacles or flesh should be firm and intact.
3. There should be no discolouration.
4. The shellfish should have a pleasant fresh sea smell.

Storage of fish

1. All fish should be scaled, gutted and cleaned before storage.
2. Fish maybe stored in a refrigerator for three days or in the freezer for up to six months.
3. All fish should be tightly wrapped in plastic film to prevent drying out and the transference of the fish smell to other foods.
4. Once fish is thawed it should not be refrozen.
5. Smoked fish is best wrapped in foil to prevent sweating.
6. When freezing fish fillets/cutlets it is best to wrap each piece individually.

Preparation of seafood

1. Fresh seafood is best prepared on the day of purchase. All seafood is delicate and must be cooked gently and not overcooked. Seafood is often basted to prevent drying out.
2. Crustaceans will change colour to red, pink or orange when cooked.
3. Molluscs such as mussels open their shells when cooked. Do not eat those that will not open, discard them.
4. Fish maybe poached, fried (deep and shallow), grilled, steamed, baked and microwaved or served raw.
5. To test if fish is cooked it should break into flakes when prodded.

Note: Green prawns is another term for raw prawns.

Activity 1.26: Cooking fish

Design a basting mixture suitable for use at home when barbecuing or grilling fish or other seafood. (Some suitable ingredients for a base are lemon juice, wine, oil.)

E. Milk

The main source of milk in Australia is produced from cows, however smaller quantities are obtained from goats and sheep. Sheep's milk is used in cheesemaking and there is increasing interest in the use of goat's cheese in today's cuisine. Soy, rice and coconut milk are also available, but for this purpose we will investigate cow's milk.

Milk has been available for many centuries in various forms from yoghurts to cheeses and these products are continually changing to meet market demands and trends.

Milk and milk products provide 10 essential nutrients including calcium, phosphorous, zinc, B Vitamins, protein and carbohydrate. Milk is the richest source of calcium in the Australian diet.

Did you know ...

Recent research suggests that milk may also provide antioxidants, bone growth promoting benefits and immune protection factors.

Nutritional value of milk

- **Protein:** the protein in milk is complete in that it contains all the 22 essential amino acids. The two main protein groups are: **Casein** accounts for approximately 80% of the protein. Casein is present in the curds of milk which form when acid is present or when the enzyme rennet is added. The other protein group **lactoalbumin** and **lactoglobulin** is found in the whey of milk.
- **Fat:** the fat content of milk is mainly in the form of butterfat and is present in very tiny globules which are suspended in the fluid. When milk has not been homogenised and is left to stand the fat clumps together to form cream which rises to the surface.
- **Carbohydrate:** lactose is the sugar present in milk. It has only a slightly sweet taste.
- **Minerals:** calcium is the main mineral found in milk and is essential for the growth of strong bones and teeth. As people age there is an increased risk of osteoporosis if there has been an inadequate intake of calcium over the years. This is the main contributing factor to the cause of osteoporosis. Calcium is easily absorbed by the body. Phosphorous, magnesium, potassium, sodium, chlorine and sulphur are also found in milk in varying quantities.
- **Vitamins:** the butterfat present in milk contains the fat soluble vitamins of A and D. Vitamin D is essential for the absorption of calcium and phosphorous. Milk also provides the B group vitamins of riboflavin, and smaller amounts of thiamine and niacin. These vitamins are easily destroyed by light and heat.
- **Water:** milk is made up of 87% water.

Over the years there have been many changes to milk. In the early days milk was consumed as is, with no modifications. Milk was delivered fresh daily and the housewife would scald the milk, killing any bacteria present to ensure it was safe to drink.

The milk would be then left to stand and a layer of cream would rise to the top and this was known as scalded cream. Often this cream was eaten with jam and fresh bread or scones. With technological advancement milk pasteurisation was introduced to make the milk safe to consume.

Homogenisation is a process that splits the fat globules into small particles so that they remain distributed throughout the milk which provides a consistent flavour and texture throughout.

Pasteurisation is when milk is heated to a temperature high enough to kill disease producing bacteria.

Modified milks

Whole milk has been modified in various ways to increase the health benefits of milk. The forms are described below.

- **Whole milk:** is the traditional milk with a fat content of 3.8%.
- **Reduced fat milks:** are milks which have less than 2% fat.
- **Skim milk:** is milk which has no fat content and no cholesterol. It has also the lowest Kilojoules content.
- **Added nutrients:**
  - Calcium is added to increase its content by 17%.
  - Protein, magnesium, zinc may be added to improve muscle growth and repair.
  - B group vitamins may be added for added energy.
  - Calcium and magnesium may be added for bone strength.
  - Omega-3 fats, Vitamin B6 and folate are added to improve heart health.
- **UHT (ultra heated milk):** uses a process whereby milk is heated to 135°C and held at this temperature for two or three seconds, then rapidly cooled. Milk treated by this method may be stored without refrigeration for six months.

Selection and storage of milk

- Check the use-by date when purchasing.
- The aroma should not be sour.
- It should have a smooth pouring consistency.
- Store in refrigeration at 4°C.
- Keep milk covered to prevent absorption of flavour.
- UHT MILK maybe kept in a pantry but once opened must be stored in a refrigerator and consumed within the recommended time frame.
- Milk containers should be full and not leaking.

Activity 1.27: Milk

1. Compare five different types of milk. Read the labels to determine the nutrient content and record, then evaluate the flavour, appearance and texture. Summarise your findings in a table.
2. Milk is also available as powdered, condensed and evaporated. Write a brief description of each and discuss the advantages and disadvantages of purchasing milk in these forms.

CLASSIFICATION: STAPLE FOODS

Staple foods form the basis of a traditional diet and they vary from country to country. Staple foods are usually inexpensive and are of plant origin. They are high in carbohydrate and are foods that may be stored. Staple foods may be cereals like wheat, rice, corn, rye or starchy vegetables such as potatoes and they are commonly used as part of every meal.

In Australia there are many foods that could be considered staples. Due to our different climatic zones and advanced farming techniques as well as cereals being a staple of Australia, meat such as lamb and beef could also be considered staples. Seafood and vegetables are also included as staples. These foods are frequently used in the everyday diets of people in Australia.

When asking Australians what they think is a staple food of Australia the answers are varied from lamb, beef, coffee, bread, vegetables, rice and pasta to meat pies. Over the years our staple foods have changed from the use of indigenous foods (kangaroo, bush tomatoes), to foods introduced by our early settlers which had a British influence (meat which was usually lamb and three vegetables), to present day where the food we eat is influenced by our multi -cultural society. The lush pastures of Australia produce and provide a plentiful supply of beef and lamb.

Cereals are readily grown in large areas. The oceans provide a variety of fish, lobsters, scallops, oysters and mussels. An array of vegetables, melons, citrus fruit, grapes, stone fruit, berries, cherries and apples are also grown. We are fortunate to have such a wide variety of fresh foods.

Much of the information about specific staple foods has been covered earlier in this chapter.

Activity 1.28: Staple foods

What influence has the multicultural society had on staple foods in Australia?

Staple foods are generally considered to be of cereal origin and high carbohydrate foods. However, other foods are also considered staples for different cultures and may include spices, herbs, meats, vegetables and fruits. If you were investigating staple foods in the Mediterranean and Asian regions what would you anticipate finding in a kitchen pantry? What would you consider to be the staple foods found in your home pantry? Compare your results with other class members.

Factors that affect the supply of staple foods

There are many issues affecting the production of food from the very initial stages to finally being served on a plate (commonly referred to as 'paddock to plate.') Let's look at the issues which affect the quality and quantity of staple food when it is produced.

Many environmental factors influence the production of staple food and these vary from place to place and from year to year.

Soil type

The care and maintenance of soil and farming lands is important within farming practices. Excessive removal of trees by pioneers has resulted in increased soil salinity rendering it unviable for food production. The use of too many chemicals has damaged the waterways. Soil fertility is important and fertiliser use must be managed carefully to maximise production and minimise damage.

- **Chemical fertilisers** are needed to help the plants grow but in the long term may eventually damage the soil. It is important that the correct fertilisers are used to avoid as much long-term damage as possible to the soil. Soil will slowly lose its natural goodness and it will become harder to grow effective crops without fertilisers.
- **Crop rotation** helps to prevent the nutrient level decreasing and helps to maintain the quality and quantity of a crop. By cropping legumes one year nitrogen is deposited in the soil ready for cereal crops the next year.
- **Soil erosion** is another problem encountered by farmers. Wind and water erodes vital topsoil needed for crops and grazing animals. This problem is exacerbated when there are few plants holding the soil in place.

Much research is being done to develop crops that need less water and are resilient to the higher salt content in the soils in some farming areas. Soil improvers are being developed so that less water is also needed.

Pests, weeds and diseases

Crop diseases such as rust in wheat can be detrimental to the productivity of crops. Fungus, weeds and insects also have a damaging effect on the growth of crops. To prevent this damage, crops are sprayed with herbicides and pesticides which in the long term can damage the environment.

Pests can be classified in to two groups:

- insects – for example, locusts
- other – for example, rabbits, foxes, kangaroos.

Chemical herbicides and pesticides can also build up levels of toxicity in the soil. These chemicals are used so that the crops can grow to maximum potential without damage from weeds, pests and diseases. The use of these chemicals must be carefully monitored to avoid high levels of toxicity. Excessive use of herbicides and pesticides can also affect the ecology of the soil.

Water

With Western Australia's ever changing rainfall pattern, the management and use of water is crucial. Farming practices can be affected by drought and floods. Dams are used to store water during dry periods but they are reliant on being replenished with winter rains. Other disturbing concerns are water salinity and the pollution of waterways. A lack of rainfall is a serious matter as it means there is less food produced and farmers suffer financially. On the other hand, floods can wash away valuable farming topsoil and spoil crops that have been planted.

Other environmental factors

Other environmental factors that influence the production of food include:

- **Climate:** different climates are suitable for different crops and livestock.
- **Terrain:** some crops grow better in a flat landscape while other crops may be grown in any type of terrain.
- **Fire:** in the hot summer months fire can destroy and damage crops and kill livestock.
- **Hail/frost/snow:** these climatic conditions can be very damaging to crops and may even wipe out the supply of produce.
- **Cyclones/wind:** high winds will damage crops and may wipe out a season's supply of a particular food.
- **Proximity to the ocean:** this influences the amount of readily available fresh fish and seafood.
- **Salt levels in the soil:** the removal of trees and natural vegetation has resulted in the water table rising causing the concentration of salt to increase in the soil which then hinders normal plant growth.
- **Drought:** this may result in crops having a poor yield or in the worst case scenario the crops are not planted at all. The food supply is then reduced making it more expensive for the consumer.

Activity 1.29: Staple foods

1. Investigate water pollution and discuss what monitoring should be done to look after the farmer's water supply.
2. Select a range of breakfast cereals on the market and complete a comparison of the nutritional value and cost. Explain the ingredients in the various cereals.
3. Investigate three types of seafood available from Western Australian waters. Describe the characteristics of the seafood, availability, cost and appropriate methods of cooking.

ENVIRONMENTAL AND ECONOMIC CONSIDERATIONS FOR PURCHASING LOCALLY PRODUCED COMMODITIES

Local foods are those that are grown and/or produced within the immediate area that you live in. For example, if you lived in the Kimberley, local foods would be those produced north of the 26th parallel. Because Western Australia is such a vast area, for marketing purposes local foods refer to those grown in Western Australia.

A major aim of today's society is to protect the environment and to avoid using excessive amounts of natural resources. Foods which are produced locally and seasonally are often made available with limited packaging and processing. Therefore they are often referred to as being environmentally friendly.

From paddock to plate the more processing involved of the food the greater the impact there is on the environment. Energy in the form of fossil fuels is initially used by the machines when harvesting crops and is involved right through to the disposal of waste after the food has been used in the home or food industry.

Increased transport and packaging results in greater cost of the product and this is usually paid for by the consumer. In most cases this is one of the reasons why processed food is more expensive than the natural product and more damaging to the environment.

Food availability

The availability of food is dependent on many factors.

With ideal seasonal conditions food supplies will be abundant, whereas in times of extreme heat or drought the food supply becomes more scarce. The greater the supply of the food the cheaper it is.

Improved farming techniques and research has meant many new varieties of foods have been developed. This has resulted in improved flavour, appeal, yield and shelf-life, and cost factors have been reduced where possible due to improved production techniques. Improved processing techniques and methods of transportation have also resulted in food being available in some cases all year around or at least with an extended season.

Transport

Food Miles

'Food miles' is the distance food is transported from the time it is produced until it reaches the consumer. It is a term used to make consumers aware of the environmental impact and cost of food miles.

Air travel creates the most pollution; road transport is the next; whereas trains and ships/ boats create the least pollution. Food is transported from its place of production or origin to the food processing plant. Packaging materials are transported to the processing plant Transport is then needed to transfer the food product to the retailer and then to take the food home. Once the food is consumed, transport is then needed again to take the waste away.

The further food is transported the greater the amount of pollution (carbon dioxide) that is produced. This involves all points from growing to place of consumption. The greater the transport needs, the higher the cost to the consumer and the environment.

Packaging

Packaging does help to protect and preserve the food as it is transported and presented for sale. In most cases, if food was not packaged it could become damaged, made unsafe to eat and its shelf life could be reduced. Therefore, packaging is important to preserve the quality of food and prevent undue waste. The question is the degree and type of packaging that should be used. Energy and resources are used in the making of packaging by the manufacturers; fuels are burnt which then pollute the environment.

After use, food packaging becomes waste and has to be transported from the place of use to waste disposal areas. Some forms of packaging harm the environment as they do not rot or disappear, that is they are non-biodegradable, and the waste piles just get bigger. With recent developments some packaging has become biodegradable and other packaging is recycled for other uses.

Natural resources are used to make packaging. For example, metal is used for tins and cans, oil is used in the making of plastics and wood is used for making paper and cardboard.

Activity 1.30: Environmental considerations

1. Name some products that are made from recycled materials.
2. Is all packaging really necessary? What foods are packaged unnecessarily?
3. In the olden days food was available mainly in bulk. Why did it become popular for food to be sold in packages?
4. Why are farmer's markets considered to be environmentally friendly?
5. How could you reduce your effect on the environment when growing food and purchasing food?

Food wastage

Not only is food wasted when it is not used, it is also a waste of resources that were used to grow and process the food. Does it really matter that the apple has a blemish on the skin or the pear is misshapen? Consumers should purchase food wisely to prevent wastage from food deterioration. Food should be stored correctly to prevent wastage.

Food costs

The two main economic factors are the purchasing price of food and the money available to spend on food. Low income does not mean a less nutritious diet however it may mean limited food choices.

Food costs may be influenced by fuel and transport costs. The higher the fuel costs the higher the food costs, and if the costs become too high, the particular food product becomes out of reach to a large number of consumers.

Often if there is a glut of a product (usually a variety of fruit or vegetable) food prices will be reduced. Conversely if a food is in short supply (often due to damage from storms, drought) it will be higher in price. At these times of scarcity the food may be shipped in from other parts of Australia or overseas. This becomes more expensive because of the cost of transport and agricultural policies related to imports and exports.

When foods are not in season it is often cheaper to purchase them in a processed form such as canned or frozen. However, the raw foods are usually cheaper than their counterpart processed foods. Frequently, the more ingredients on the label the higher the price will be.

Activity 1.31: Cost comparison

Conduct a cost comparison of the following foods in the natural and processed form then do three of your own choice.

Natural Food	Cost per Kilo	Processed Food	Cost per Kilo
Grapes		Sultanas	
Apples		Canned apples	
Corn		Frozen corn on the cob	
Fresh chicken breast		Chicken breast roll	
Blueberries		Frozen blueberries	

When purchasing food, most of the food budget should be spent on fruit, vegetables, legumes, bread and plain cereal foods such as rice, pasta and plain breakfast cereals. Moderate amounts of money should be spent on meats, poultry, fish and milk products. The lowest amount of money should be spent on the least nutritious foods. These are foods which are high in sugar, fat and salt, for example, cakes, biscuits, chocolates, cordial and soft drinks.

Choosing to purchase locally produced foods has many environmental and economic advantages. For example:

- Food is fresher.
- Food is cheaper.
- Less packaging is needed.
- Transport costs are reduced considerably.

CHAPTER 2
Properties of Food

Key Concepts

- **Sensory properties**
- **Vocabulary for evaluating food**
- **Physical properties of staple foods**
- **Physical properties of processed foods**
- **Selecting processed foods**

INTRODUCTION

All foods have their own characteristic set of physical, chemical and sensory properties. Physical properties influence our selection of food. A tasty rockmelon should smell ripe and have no soft spots on the skin. A ripe avocado should be soft around the stalk end, but if the avocado is hard it is not ready for eating. When buying strawberries you look for bright red, firm berries with bright green leaves and a sweet aroma.

During food preparation the properties of food change to make the food often more appealing in taste, texture and aroma. The appearance of food will also change, although not always. When cooked, the texture of a potato becomes soft and more easily digestible. Some people prefer the flavour and aroma of cooked onions rather than the raw product. The texture of carrots becomes more tender when cooked rather than crisp and crunchy when raw. All of these are examples of beneficial changes to the food's physical properties.

When selecting foods and determining how they are going to be prepared or cooked you use sensory evaluation to assess the food's qualities.

The physical properties of food may change during storage and for this reason it is important that food is stored correctly to maintain optimum quality and peak physical properties. A ripe banana will soon become soft, and develop brown blotches on the skin and a more intense flavour if stored incorrectly. A lettuce will lose its crispness if not kept refrigerated. Milk will become sour very quickly if it is not kept chilled and the texture will also take on a curdled appearance.

SENSORY PROPERTIES

All of your senses are involved when you eat an item of food. These are outlined below.

Sensory evaluation is when you use your senses to make a judgement about food.

1. Flavour

Flavour is a combination of aroma and taste. Flavour refers to whether a food is generally sweet, sour, salty or bitter.

These tastes are found in different areas of the tongue. Umami is another category of taste which is sometimes used. It is a Japanese word meaning pleasant to the taste, agreeable, good, mild, savoury and delicious. Umami is generally considered as a savoury/meaty taste and is often associated with Asian food. The longer a person is exposed to a particular taste the less pronounced it becomes as your taste buds adapt over time. Some tastes will overpower other tastes such as spicy over sweet.

Various factors can affect and alter the taste sensation:

- **Temperature:** the sweet sensation is intensified at a higher temperature whereas the same amount of salt is less salty at a lower temperature. The bitter taste is less intense at hot or cold temperatures.
- **Sweet sensation:** the sweet sensation modifies the sour, bitter and salt sensation.
- **Consistency of foods:** the same quantity of sugar set to a different consistency gives a different degree of sweetness.

Activity 2.1: Flavour

1. Can you think of another situation where one taste overpowers other tastes?
2. People are sensitive to different intensities of tastes. An example of this is when people have different tolerances to chilli. What one person considers hot another may consider to be mild. What tastes do you have a low tolerance for (that is cannot eat too much of)?

2. Aroma

Food must have an acceptable aroma for people to have any desire to consume it. A pleasing odour makes food enticing for people whilst an unpleasant odour conjures up negative thoughts about this food. Warm and hot food has a more distinctive aroma compared to cold or frozen food. For example, ice cream has very little aroma.

People have the ability to smell hundreds of compounds. They have individual differences in their ability to smell certain compounds and have different sensitivities to different smells.

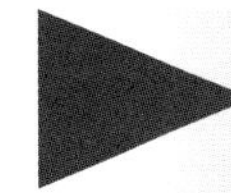

Activity 2.2: Aroma

1. Which smells do you like?
2. Which smells do you dislike?

People can have diminished smell ability when they are sick, especially with colds when the nasal passages are blocked. People have the ability to adapt to a strong smell. This occurs when you detect a strong smell then, after a while, you no longer notice the smell, for example, workers in the fish industry.

Due to the chemical compounds in food the aromas can change with changes of temperature, for example onions. Products that have started to deteriorate, for example sour milk, can also be determined by smell. This is a very handy mechanism to prevent people eating unsafe food. Often if a food smells good it tastes good and if a food smells bad, it may not be safe to eat.

3. Appearance

The appearance of food is usually the first aspect or quality of food that people judge. The appearance of food will tell you whether it is fresh or stale, damaged or in peak condition, prepared with care, and cooked to the correct degree. It involves the colour, shape and form of the food.

Certain colours are associated with flavours. The colour red often identifies a strawberry flavour. A well-cooked cake is expected to have an even brown colour on the crust. If custard has lumps in it, before tasting the food a person will have predetermined that the texture is not suitable for this food. Milk which has a curdled appearance shows that it has gone sour. The interior of a cake can indicate whether a cake is over or undercooked. Raw or gluggy mixture indicates undercooking whereas a very dry crumbly appearance indicates overcooking. If food is not visually appealing, people are unlikely to eat it.

Activity 2.3: Appearance of foods

When choosing the following foods describe what you would look for in the appearance so as to obtain prime quality.

- Mince
- Tomatoes
- Bread
- Potatoes
- Pears
- Strawberries
- Grapes

4. Texture

This describes how food feels when touched and handled, and also the feel in the mouth. An acceptable texture is important if a person is going to choose a particular food again. When the physical properties of food are altered the texture is often changed. For example, apples are crisp and crunchy when raw however when they are cooked they have a soft texture.

Detecting the texture of food is done in the following ways:

- **Smoothness/grittiness** is determined by placing food between the tongue and the roof of the mouth.
- **Crispness/crunchiness** is determined by the pressure used by the teeth to bite the food.
- **Chewiness/toughness** is determined by the amount of work the teeth have to do before the food is swallowed.

Activity 2.4: Texture of foods

Describe the texture changes to the following foods.

- Poached egg – hard boiled egg
- Spinach – steamed spinach
- Fresh pear – canned pear
- Fresh bread – toasted bread
- Cream – whipped cream
- Fresh berries – frozen berries

5. Sound

Certain sounds are associated with particular foods. Hearing enhances both the anticipation and the enjoyment of eating food.

Activity 2.5: Sounds of foods

1. Specific sounds are associated with particular foods. What foods do you associate with the following sounds?
   - Sizzle
   - Crunchy
   - Popping
   - Slurp
2. Can you think of some other sounds associated with particular foods?

DESCRIPTIVE VOCABULARY FOR EVALUATING FOODS

Here is some descriptive vocabulary that you may use when evaluating foods.

bitter
brittle
chewy
clear
coarse
crisp
curdled
dull
fine
firm
flat
frothy
granulated
limp
lumpy
moist
rubbery
smooth
sparkling
stale
sticky
stiff
stringy
syrupy
tasteless
tender
tough
watery

Activity 2.6: Vocabulary

How many more words can you add to the list?

Fruit and vegetables

Property	Description
Appearance	Fruit and vegetables come in many shapes and sizes. The colours of vegetables are mainly green, yellow/orange, red/blue/purple or white/pale and creamy.
Flavour	Fruit and vegetables may be sweet, sour, subtle or bland. Flavours develop and change during the cooking process and according to degree of ripeness.
Texture	Fruit and vegetables vary from crisp, firm to soft.
Aroma	Many vegetables have a strong aroma due to the sulphur compounds found in them. The aroma will change with different cooking processes and by adding different substances to the cooking process. • Fruit generally has a more pleasant aroma. • Many fruits may be identified by their aroma.

Changes of colour do occur in the cooking process of fruit and vegetables therefore careful management is needed.

- **Green** colours may become dull with a greyish tinge. In the past people would often put a pinch of baking soda in the water to preserve the bright green colour. However this practice has not continued as the texture of the vegetables became mushy.

What could you do to maintain the bright green colour during the cooking process?

- **Yellow and orange** colours change very little in the cooking process.
- **Red, purple and blue** colours usually retain their colour when cooked.
- **White** colours show very little change.

Activity 2.7: Textures of fruit and vegetables

Give examples for the following statements.

1. When cooked the cell walls of the vegetables change and they become softer.
2. During the cooking process some of the water content of the vegetables is lost and it takes on a wilted appearance.
3. Starchy vegetables absorb water and this increases the tenderness.
4. During the cooking process water passes through the cell membrane of fruit causing them to swell and burst. Fruit may lose their shape and become pulpy.

Did you know ...

Baking and frying partly caramelises food when there is greater exposure to heat, in comparison vegetables which are boiled have a milder flavour. Is this why people prefer roast potato?

Caramelisation occurs when the sugar in food is browned. Flavour changes also occur.

Activity 2.8: Shape and texture

1. For each of the following cooking methods discuss the effect on the shape and texture of the fruit and vegetables.
   - Poached/stewed fruit
   - Baked vegetables
   - Fried vegetables
   - Boiled spinach
   - Steamed vegetables
2. What role does sugar play when fruit is cooked in a syrup?

Enzymes

Enzymes are the catalyst that speed up a chemical reaction. Enzymes are present in fruit and vegetables and are responsible for the ripening process which continues after picking. Too much enzymatic action will cause damage (overripening) to the fruit and vegetables affecting both the flavour and texture. Refrigeration will slow down the enzyme action, hence the recommendation that most fruit and vegetables are best stored in the fridge. Heating will cause the destruction of the enzyme, therefore vegetables are blanched before freezing to reduce or minimise their destruction.

Sensory evaluation

Let's look at an apple to see what you would expect with a sensory evaluation.

- **Appearance:** red or green in colour or a combination of both (or yellow if it is a golden delicious)
  - smooth skin
  - rounded shape
  - no blemishes.
- **Feel/Texture:** firm and smooth to touch
  - crisp and juicy when eaten.
- **Smell:** sweet fresh.
- **Sound:** crunchy sound when eating.
- **Taste:** sweet with a tang sensation.

Activity 2.9: Sensory experiences

1. Complete the following chart describing the sensory experiences you would expect with the following food items. Try not to repeat any of the descriptive words used. Use the previous example of an apple to help you complete this activity.

Food	Appearance	Smell	Texture/feel	Taste	Sound
Potato crisps					
Fresh wholemeal bread					
Milk					
Chocolate					

Food	Appearance	Smell	Texture/feel	Taste	Sound
Rockmelon					
Grapes					
Ice cream					
Celery					
Tomato					

2. Students are to work in pairs with one student to be blindfolded and the other to give the blindfolded student samples of foods and/or drinks and see how many they identify correctly. Between each sampling the student should rinse their mouth with water. Students should then swap roles. Record results in a table such as the one below.

Product	Aroma	Flavour	Right or Wrong

3. How accurately did you choose the correct products?
4. Did the smell of the food influence your final decision?

Meat

Quality	Description
Appearance	Beef should be bright red in colour, lamb should be a deep pink colour, pork should be pale pink and poultry should be pale pink if it is from the breast and slightly darker pink if it is from the thigh or leg.
Flavour	Meat has distinctive flavours depending on the type of animal and it may also vary with the mixture of feed used. Chicken has a mild flavour whereas the flavour of fish depends on its variety. Some fish have a mild flavour while other types are strong.
Texture	Meat and fish should be moist and pliable.
Aroma	Fish should have an 'ocean' smell.

During the cooking process the protein in meat coagulates. The cooking process softens the muscles fibres, making it more edible and easier to chew and develops the flavour.

Activity 2.10: Meat

1. Describe marbling.

2. Compare pieces of steak that are purchased as veal, yearling beef and prime beef. Explain how the appearance differs in the colour of the meat, the colour of the fat and the texture of each. Repeat this process comparing a lamb chop to a mutton chop.

Milk

Quality	Description
Appearance	White liquid, with the creamy colour intensifying as the cream content increases
Flavour	Bland
Texture	Smooth, the viscosity varies as the milk is processed in different forms
Aroma	Slight creamy odour

Any changes to the mild flavour of milk are easily detected. Dried and evaporated milk have a more intense flavour. Fresh milk has a faint creamy aroma but once it turns sour it develops a strong, acidic smell.

PHYSICAL PROPERTIES OF STAPLE FOODS

Let us look at the raw food groups and their physical properties and how this affects their selection and use. Changes occur through the use of different equipment, cooking methods and the combination of different ingredients.

Remember the physical properties of food refer to its colour, size, shape, volume, viscosity and texture.

Physical properties of cereals

- They are usually white in colour.
- They are bland in flavour.
- Starches will form cloudy, white or transparent gels.
- When dry heat is applied to a starch it will brown and also change in flavour.
- When moist heat is applied to a starch the granules will swell, burst and thicken a liquid.
- Grains such as rice absorb liquid in the cooking process causing the grains to double or triple their size.

When subjected to heat, starches can change from their usual white colour to different shades of brown depending on their use. The degree of darkness of the brown colour affects the flavour and different shades are acceptable for different products. An example is when some people prefer lightly cooked toast while others prefer a darker shade, and when baking cakes and biscuits a darker shade may indicate the product is overcooked. When a cereal is browned it develops a sweeter taste.

When processing starch products, some important processes occur.

- **Dextrinisation:** this is the breakdown of starch during baking, which adds sweetness to products. Dextrins are the compounds which form during the breakdown of starch. They form during the browning of toast, browning of cakes, bread and biscuits, and give food its cooked taste. If the colour is too brown a burnt taste will develop. The crust is initially crisp but becomes soft when exposed to moisture.
- **Gelatinisation:** this is the thickening property of starches when heated in a liquid. The starch cells swell and absorb the liquid. When the starch grains are heated they gel. This process is important in the making of sauces and gravies. It is known as gelatinisation. In the cooking process the cellulose is broken down making it easier to digest. The final thickness of a sauce or gel will depend on the proportion of starch to liquid, that is increased liquid creates a thinner consistency.

Did you know ...

When a gel cools often it will lose liquid. This is called weeping or syneresis.

Starches in food preparation

- **Wheat flours** have a nutty, mild flavour when exposed to heat. They are used to thicken sauces and provide texture.
- **Cornflour** helps absorb moisture in products such as icing sugar. It is often used as a dusting powder on confectionery such as Turkish Delight. Cornflour is also used to thicken sauces.

- Arrowroot gives a clear smooth appearance when used as a glaze on fruit-topped tarts.

Activity 2.11: Starch

1. Place 1 tablespoon of a selected starch (eg. cornflour) in a small saucepan and gradually mix in 1 cup of water then place over heat stirring all the time until the mixture boils. Record the degree of thickness compared to the other starches being used. Then place in a cup (or use other similar size containers) in the fridge. Leave in the fridge until cold. Use a numbering system to record the degree of thickness. For example, 5 = very thick through to 1 = very thin. Record your results in a table such as the one below.

Results:

Starch	Colour after boiling	Degree of thickness after boiling	Degree of thickness after refrigeration	Use in food preparation
Semolina				
Cornflour				
Plain flour				
Rice flour				
Arrowroot				
Wholemeal plain flour				
Bread flour				
Place white plain flour in an oven until it turns golden then proceed to mix with water and boil				
1 tablespoon flour and 3 teaspoons sugar				
1 tablespoon flour and add 3 Tablespoons lemon juice to the water				

2. Starches are used extensively in the food industry and in food products. Research and discuss the various uses of starches in the food industry and in what products they are found. Read food labels to find the starches that are most commonly used and answer the following questions.
   a. Which starches used are used in the food industry?
   b. What is the most commonly used starch? Why?
   c. What products are starches are found in?
   d. Discuss the role of starches in the food industry.
3. Select a range of rice grains (arborio, long grain, short grain, jasmine, brown) and cook a sample according to the directions on the packet. Once cooked compare with the uncooked product. Can you detect flavour differences?
   a. Which product increased in size the most?
   b. Which product increased the least?
   c. Why is this knowledge important to know in food preparation?
4. Tapioca and sago are less commonly used starches these days. Where are they from? What is their appearance like? Find a recipe (Hint: tapioca puddings are an old-fashioned favourite) and try it out in class and conduct a class tasting. Discuss your opinions.
5. Pasta is a very popular food. There are many different shapes which have various uses. Draw an example of the following pastas and list their most appropriate use.
   - Spaghetti
   - Lasagne
   - Macaroni
   - Risoni
   - Penne
   - Fettuccine
   - Cannelloni

Physical properties that influence selection and use of raw and processed food

Colour, shape, size, volume, viscosity, colour and texture are the physical properties of food . All foods have particular physical properties that change when the food is cooked or processed.

Size

The size of fruit and vegetables can affect their texture. A very large carrot may be considered to have a tough/woody texture whereas a small carrot is crisp and juicy.

The size of fruit and vegetable pieces can also determine the method of cooking. For example, small potatoes are usually cooked whole while a large potato is cut into pieces for cooking.

When choosing pears for poaching, whole fruit is chosen that is of even size. Often the smaller the fruit the sweeter the flavour. The size of a piece of food can affect the cooking time. A much longer period of time has to be allowed to cook large cuts of meat while smaller single serve pieces can be cooked very quickly.

A piece meat with large muscle fibres indicates toughness, while short muscle fibres indicate tenderness.

Shape

The shape of a piece of fruit is important for display purposes, particularly when serving something like whole poached pears or using whole strawberries for a decoration.

Different varieties of pears, pumpkin and potatoes all have a different shape and this helps to determine each particular variety. The shape of a piece of meat, for example a lamb cutlet, is also important when serving for aesthetic reasons.

When purchasing processed food such as frozen potato chips you would expect them to be all the same shape and size.

Colour

The colours of vegetables are mainly green, orange/yellow, red/blue/purple or white/cream. The colour of a piece of fruit or vegetable will often help determine how ripe it is.

When purchasing processed food, if the colour is quite different to the natural colour consumers find it off-putting as it appears artificial.

The colour of the skin of an apple can determine the variety of apple it is, for example green skin indicates a Granny Smith apple while a red/green skin indicates possibly a Pink Lady apple. This also applies to pumpkin.

A dark amber coloured olive oil indicates it is the extra virgin variety.

When purchasing beef it should be bright red in colour, lamb should be a deep pink, pork pale pink and poultry should be pale pink if it is from the breast and slightly darker if it is from the thigh or leg.

Marbling of beef indicates tenderness and dark creamy coloured fat indicates the meat is from an older animal.

Milk is considered to be a white liquid with the creamy colour intensifying as the cream content increases. A creamier colour also indicates if the milk has been processed and some of the water content reduced.

Homogenisation is the process where the fat globules are split into small particles and are suspended throughout milk. If milk has not been homogenised, a layer of cream will rise to the top of the milk on standing. Homogenised milk has a consistent flavour and colour throughout.

Volume

When purchasing eggs, the volume is very important. When egg whites are whisked they form a foam. For example, when making products which require aeration (sponge cakes, meringues) from egg whites or serving eggs as a meal it is better to choose a larger size egg.

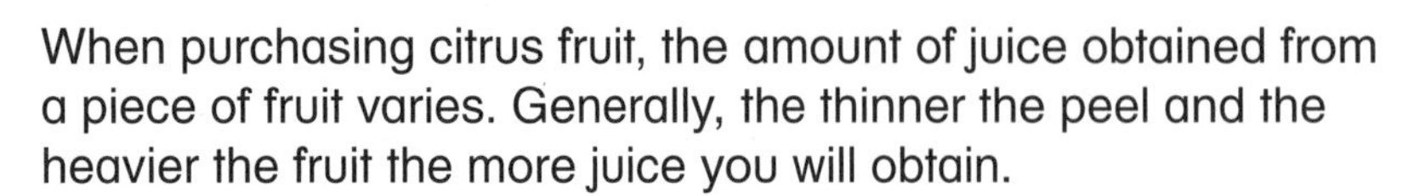

When purchasing citrus fruit, the amount of juice obtained from a piece of fruit varies. Generally, the thinner the peel and the heavier the fruit the more juice you will obtain.

Viscosity

The viscosity of milk changes once some of the water has been removed, as in evaporated and condensed milk where the viscosity increases. This has benefits for use in the cooking of sauces and desserts where a thicker consistency than straight milk is required.

When purchasing yoghurt, the viscosity varies with particular brands and types of yoghurt. Some yoghurt will need to be stirred before use to thoroughly mix any of the product that has separated.

Jam, margarines and spreads have to be of the correct viscosity to be easily spreadable.

The viscosity of egg whites determines their freshness. The thicker the egg white the fresher the egg.

Did you know ...

Did you know that sugar stabilises egg whites making them more structurally sound and lasting for longer periods of time.

The viscosity of fats and oils increases when heated.

Activity 2.12: Cooking and changes in physical properties

Let's investigate the changes to the physical properties of fruit and vegetables when cooked. Do a sensory evaluation of colour, flavour, aroma and texture of six different fruit and vegetables of your choice. Remember to cook your fruits and vegetables carefully to maintain optimum quality and avoid overcooking as this will be detrimental to your results. Record your results in a table such as the one below.

Fruit/ Vegetable	Colour fresh	Colour cooked	Texture fresh	Texture cooked	Aroma fresh	Aroma cooked	Taste fresh	Taste cooked

1. Which foods have developed a sweeter taste? Why?

2. Which foods had the most change in aroma?
3. Which foods had the most change in texture?
4. Which foods had the least texture change?
5. Which colour changed the most?
6. Which colour showed little change?
7. What are your overall conclusions?
8. List your recommendations for cooking vegetables so as to maintain colour, flavour and texture.

PHYSICAL PROPERTIES OF PROCESSED FOODS

When purchasing processed foods their physical properties assist to determine the quality. Many of the foods which have been dried, canned or frozen can contribute favourably to the flavour of a meal. The many uses of processed food contribute a lot of variety to a person's diet.

The processed food should be as similar to the fresh version as possible. With technology there is a wide range of processed foods available.

Activity 2.13: Processed foods

Just think about all the ways we can purchase pineapples. They may be fresh, dried, canned or juiced. Can you think of any other ways?

Selecting processed foods

1. Dried fruits should be plump, tender and of a good colour.
2. Ice cream should have no sign of ice crystals, indicating some degree of thawing has occurred.
3. Cheeses should have no sign of package damage or mould.
4. Canned food should not be dented as this could cause a break in the seals causing food spoilage. There should be no bulging in the ends of the tins.
5. When buying dried foods check for insect infestation where possible.
6. Dried foods should have no sign of grit or dirt.
7. Dried herbs should have a fresh distinctive aroma and be of a good colour. Once this has gone the herbs will lack flavour.
8. Frozen fish should be wrapped in separate pieces.

9. Frozen vegetables should be free flowing and not clumped together as this is an indication of them being thawed and refrozen.
10. Frozen chicken should have no frozen meat juices around the outside as this indicates thawing and refreezing.
11. Packaging should be tightly sealed with no sign of damage.

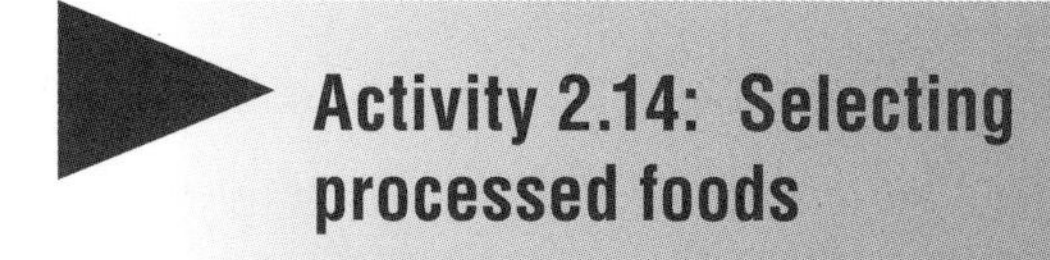

Activity 2.14: Selecting processed foods

What other points can you think of to add to this list?

Did you know ...

Lettuces can only be bought fresh.

Activity 2.15: Food evaluation

1. Frozen berries are very popular. During freezing they tend to become softer and do not hold their shape when thawed. Because of this, they are often incorporated into cooking rather than served on their own. How may frozen berries be incorporated into food preparation?
2. Do a sensory evaluation of peas completing the following chart and draw your conclusions as to the value of each type. You may repeat this exercise using another food such as potatoes or tomatoes.

Type of pea	Appearance	Texture	Flavour	Comments
Fresh peas				
Canned peas				
Frozen peas				
Dried peas				

3. Prepare a healthy dinner based on processed food. You may include pasta, tinned fish and a range of frozen vegetables; see how creative you can be. You may wish to use some of your own ideas. Compare your results to other class members. What improvements could be made? Write down your recipe and add your evaluation. When would you find it convenient to make this meal?

EXTENSION

Extension Activities

1. You are off on a **camping weekend** and will spend much of the time trekking and carrying supplies in a backpack. In this situation to carry fresh food would be cumbersome and the food would be easily damaged. Imagine a squishy banana after a day trekking, dried banana would certainly be a better choice.

   Plan a two-course evening meal at the campsite. Next to each item of food you select list the points to check before purchasing, for example, Mountain Bread should have no signs of mould, check the use by date. What breakfast foods and snacks would you take?

2. Take a tin of peaches and develop a **café standard dessert**.

3. **Frozen stir-fry vegetables** are very popular as an alternative to fresh vegetables for young people living on their own. Develop a recipe incorporating these vegetables into a one-bowl meal. Think further than a stir-fry.

4. **Your parents** are out and you are caring for your younger brother. You have the basic ingredients for a **frittata** but wish to add some interest to it. What ingredients in the freezer or cupboard could you find to add to your frittata? Prepare and evaluate your product.

5. **Your grandparents** are finding it harder to get to the shops. What foods do you suggest they keep in their **freezer and cupboard** as an acceptable alternative to the fresh version? Give reasons for your choices.

6. Write a **promotional article** for a fruit and vegetable processing company extolling the advantages of frozen fruit and vegetables.

7. In the hospitality industry the statement **'People eat with their eyes'** is often used. Explain this statement and give examples.

8. Give some examples of a healthy daily eating pattern.

9. You have been asked to prepare an article for a brochure to give to new parents regarding dental caries. Discuss the meaning of 'hidden sugars', list foods which should be avoided and include acceptable food choices. Include as many points of interest that you can think of. Give some examples of a healthy daily eating pattern.

10. Under the headings Dietary Concerns, Culture, Special Occasions, Availability of food, Advertising, Weather, Budget, Time and Skill discuss your own family's eating habits. Are there any areas where there should be changes and what advice would you give?

11. Find, modify or develop your own recipes for snack foods suitable for school that are low in sugar and 'teeth friendly'. Trial these recipes and evaluate their suitability for teenagers.

12. Write down all that you ate in the last three days. Categorise the food into two columns, essential and non-essential. Examine closely the foods that are non-essential. What main ingredients are you gaining from these foods? List them, and then describe the possible health problems that are associated with eating these foods and the effect these nutrients will have long term. Also look at possible deficiencies within the diet and problems associated with this.

13. A lot of health issues can be overcome by paying special attention to your diet. Research your family's background and take note of any health issues that may exist, for example cancer, heart disease, diabetes. Ensure to include your grandparents, uncles and aunties as well as immediate family. If there are any problems, describe how you can adapt your diet to help avoid having these potential health issues. If not, describe how you can avoid obesity.

CHAPTER 3
Nutrition

Key Concepts

- Introduction to nutrients
- Macronutrients
- Micronutrients
- Macronutrient and micronutrient requirements
- Food Models
- Foods to promote health
- Nutrition related health problems
- Factors affecting nutrition related health issues
- Modifying eating habits
- Nutritional requirements for different age groups

INTRODUCTION

Consider the statement 'You are what you eat'. Generally excess food intake can lead to a person becoming overweight while, conversely, inadequate food intake may cause a person to be underweight and malnourished. The consumption of less nutritious foods may also be a contributing factor to numerous health issues.

Nutrition refers to the process whereby the body takes in food and sustenance for the purpose of growth, health and improving life.

People have different nutritional requirements throughout the life cycle. It is important for a person's good health that they eat a well-balanced, healthy diet.

It is also important that people have a basic knowledge of nutritional issues and their implications for them, so that they can apply this to their own food needs. Good eating habits are started early in life and are important for growth (especially in younger people), for repair to damaged body tissues, for energy and to regulate body processes.

Eating well

To maintain good health people should focus on selecting a balanced diet daily rather than spoiling the diet with 'junk food' or empty kilojoule type food. A wide variety of foods should be chosen from cereals, fruit, vegetables, lean meat, healthy oils and low fat dairy products.

It is important that a person eats breakfast every day as this discourages snacking on junk food; provides energy for the body to perform to its optimum; and breaks the fast from the night before.

Snacks are also a temptation to eat less nutritious food. Many popular snacks are high in fat, sugar and salt, and low in fibre, like crisps and most bought biscuits.

Activity 3.1: Nutritious breakfast

What are some quick nutritious breakfast ideas?

Water is very important to our body. People should avoid the urge to consume too many high kilojoule drinks such as soft drinks and cordial which may contribute to obesity and dental caries. Water is a healthier alternative.

Low fat cooking methods such as grilling and stir-frying are preferred choices for everyday meals rather than the high fat methods such as deep-frying and shallow-frying.

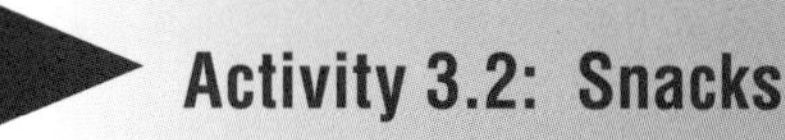

Activity 3.2: Snacks

What are some preferred choices for snacks rather than the high kilojoule snacks often consumed?

Activity 3.3: Reducing fat

How could you reduce the fat content in a roast meal?

As well as maintaining a healthy food intake a person must also exercise regularly even if it is just a 30-minute daily walk.

Activity 3.4: Daily exercise

How do you get your 30 minutes exercise each day?

NUTRIENTS

Food is the substance that makes our body work, provides the body with energy, promotes growth, assists to repair damage and protects against disease. **Nutrients** are the chemical substances found in foods which are essential to life and maintaining good health. The nutrients found in food are carbohydrates, fats, minerals, proteins and vitamins. A person's actual nutrient intake varies with age, gender, activity and size.

The main nutrients are:

- carbohydrates
- vitamins
- protein
- minerals
- fats
- water.

Nutrient	Function	Food Source
Carbohydrates	• provide energy	Bread, cereals, rice, pasta
Protein	• build and repair body cells • are important during periods of growth	Meat, fish, nuts, eggs, milk, cheese
Fats	• provide energy	Oils, fats, butter, margarine, cream
Vitamins	• essential for normal growth and development • are essential for many of the chemical reactions in the body	Fruit, vegetables
Minerals	• regulate body processes.	Wide range of cereals, meat, fruit and vegetables
Water	• forms the bulk of body fluids and secretions • moistens the air we breathe in • Is a solvent for waste products	Glasses of water, soup, fruit juices

- **Macronutrients:** are required in large amounts for healthy growth and to provide energy for the body's daily needs. The macronutrients are fats, protein and carbohydrates.
- **Micronutrients:** are found in small amounts in food. The micronutrients required by the body are minerals and vitamins. They play an important role in body functions but do not provide energy. Micronutrients are important in the basal metabolic processes (for example enzyme reactions), functioning of the heart and nervous system, and helping to manufacture antibodies to fight infection.

Basal metabolic rate (BMR) refers to the energy needed to perform the involuntary energy needs of a person's body. The BMR is highest in young people and decreases with age.

Activity 3.5: Energy needs

What are the involuntary energy needs in the body? (For example, heart beat.)

MACRONUTRIENTS

A. Lipids

Fats are solid at room temperature and of animal origin, for example butter; while oils are liquid at room temperature and are of plant origin, for example canola oil, olive oil. Fats are made up of glycerol and fatty acids. Fats and oils may be classified as follows.

1. Polyunsaturated fats

These fats have been found to reduce the risk of heart disease when they are used to replace saturated fats in the diet. An example of this would be changing from butter to polyunsaturated margarine or using olive oil rather than dripping. Omega-3 fats are also known as fish oils as they are found in oily deep sea fish. It has been found that these fats reduce the risk of heart disease therefore they have been added to products such as bread and milk.

Activity 3.6: Omega-3 fats

Name some products that have been fortified by the addition of omega-3 fats.

2. Saturated fats

These fats are derived from animals and plants and are usually associated with increased risk of heart disease.

Examples of saturated fats include butter, cream, the fat on meat and skin on chicken, while those of plant origin include cooking margarine, palm oil and coconut oil.

3. Monounsaturated fats

These are found in both olive oil and canola oil and margarines made from them as well as some nuts, seeds, olives and avocados. These fats lower the levels of bad cholesterol in the blood when they replace saturated fats.

4. Trans-fats

Trans-fats are types of unsaturated fats that, because of their chemical structure, behave more like a saturated fat in the body and in food processing.

Trans-fats may be found naturally in dairy products and meat. However they are also produced during the process of hydrogenation. Hydrogenation is the process where liquid vegetable oils are converted to semi-solid fats.

Trans-fats are often found in bought cakes, biscuits, pastries and other takeaway food. A high intake of these types of fats is not recommended.

5. Cholesterol

Cholesterol is a waxy substance and is needed by the body in small amounts for the manufacture of bile acid and in the production of vitamin D. It is found in the cells protecting the brain and the nervous system. The body will manufacture enough cholesterol for its own needs. Saturated fats in the diet will also provide cholesterol. Excessive amounts form deposits in the arteries which can lead to cardiovascular problems when the arteries become narrower and this may eventually lead to heart attacks and strokes.

Functions of lipids

- Provide the body with energy
- Act as insulation below the surface of the skin
- Protect bones and body organs
- Maintain the structure and health of the body's cells

Activity 3.7: Fats

1. Why is a very low-fat diet not recommended?
2. Compile a checklist headed 'Limiting fat intake in the diet'.

B. Protein

Protein is known as either complete or incomplete protein
and is made up of the elements of carbon, hydrogen, oxygen and nitrogen.

Our bodies require 20 amino acids, eight of which are considered essential and must be obtained from foods rich in protein. The remaining 12 are considered non-essential but are easily obtained from food or the body can synthesize them as needed.

Complete protein foods provide eight of the essential amino acids needed from food as the body cannot manufacture them. Complete protein is of animal origin and is found in meat, poultry, fish, milk, cheese and yoghurt.

Incomplete protein lacks at least one of the essential amino acids. Incomplete protein is of vegetable origin and includes nuts, pulses, seeds, wholegrain cereals and vegetables.

Functions of protein

- Growth and repair of cells
- Repair of hard cells for example, bones and teeth) and soft cells (muscle)

- Manufacture of enzymes, hormones and haemoglobin
- An energy source if insufficient carbohydrate is eaten

Activity 3.8: Protein

1. Some complete protein foods are:
2. Some incomplete protein foods are:
3. Which plant foods are high in protein?
4. How would a vegan avoid having a deficiency of amino acids? Remember a vegan does not eat meat, fish, eggs or dairy products.

C. Carbohydrates

Carbohydrates are compounds of hydrogen, carbon and oxygen and there are two main types:

1. **Simple sugars:**
   - monosaccharides – glucose, fructose, galactose
   - disaccharides – sucrose, lactose, maltose.
2. **Complex sugars:** polysaccharides – cellulose, starch.

Activity 3.9: Carbohydrates

Which foods provide the following carbohydrates?

- Glucose
- Fructose
- Galactose
- Sucrose
- Lactose
- Maltose
- Cellulose
- Starch

Functions of carbohydrates

- Important source of energy
- Important source of fibre

Some foods which are high in carbohydrate are bread, breakfast cereals, rice, pasta and potatoes.

During digestion carbohydrates are broken down to simple sugars.

The rate at which the carbohydrate is digested varies greatly and those which are digested slowly and provide the most sustained or slow release of energy are those with a **low glycaemic index (GI).**

The glycaemic index (GI) refers to the rate at which foods release sugar (glucose) into the blood stream.

High GI foods release glucose quickly into the blood stream whereas low GI foods release glucose over a much slower period of time, which delays the feeling of hunger.

Activity 3.10: Glycaemic index and dietary fibre

1. Some foods that have a low glycaemic index (GI) are mixed grain bread and legumes. Can you think of any others?
2. Some foods with a high glycaemic index (GI) are potatoes and white bread. Can you think of any others?
3. Discuss the importance of fibre in the diet.

MICRONUTRIENTS

A. Vitamins

Vitamins are essential for a healthy body and are found in a wide variety of healthy foods. They were originally identified by letters of the alphabet but now they are also given chemical names. Vitamins have a wide range of functions in the body.

Functions of vitamins

- Cell division
- Growth
- Energy source
- Eyesight functioning
- Healing wounds
- Functioning of the nervous system
- Blood clotting
- Healthy skin
- Metabolism of protein, fat and carbohydrate
- Provide resistance to infection
- Muscle tone
- Assist in the absorption of minerals

Vitamins may be described as fat soluble and water soluble.

The **fat soluble vitamins** are:

- vitamin A (Retinol)
- Vitamin D (Cholecalciferol)
- Vitamin E (Tocopherols)
- Vitamin K (Phylloquinone).

These vitamins are found in fish oil, milk, cheese, eggs, vegetable oils, nuts and leafy green vegetables. Fat soluble vitamins may be stored in the body.

The **water soluble vitamins** are:

- Vitamin B1 (Thiamin)
- Vitamin B2 (Riboflavin)
- Vitamin B3 (Niacin)
- Vitamin B5 (Pantothenic acid)
- Vitamin B6 (Pyridoxine)
- Vitamin B12 (Cobalamin)
- Folate or folic acid (B Group)
- Biotin (B Group)
- Vitamin C (Ascorbic Acid).

Water soluble vitamins are not stored in the body therefore it is important that a wide range of healthy foods are eaten frequently to maintain their levels. Folate supplements are recommended for those considering pregnancy to reduce the risk of neural tube defects. Folate is found in green leafy vegetables, liver, wholegrain cereals and fortified breakfast cereals.

Activity 3.11: Vitamins

Complete the following:

1. Vitamin A is found in:
2. Vitamin B group vitamins are found in: dairy products, lean meat, fortified breakfast cereals and vegetables.
3. Vitamin C is found in: citrus fruit, kiwi fruits, strawberries and capsicum.
4. Vitamin D is found:
5. Vitamin E is found in:

B. Minerals

Minerals are required by the body to stay healthy and are involved in the chemical reactions in the body including metabolism, cell growth and repair, and nerve and muscle function. Most minerals are found in plant foods.

It is important that a wide variety of foods is eaten to provide the range of minerals needed by the body.

Minerals essential to good health are:

- Calcium
- Chlorine
- Iron
- Magnesium
- Phosphorous
- Potassium
- Sodium
- Sulphur
- Zinc.

Some minerals are required in much smaller amounts and are known as **Trace Minerals**.

- Cadmium
- Chromium
- Cobalt
- Copper
- Flourine
- Iodine
- Manganese
- Molybdenum
- Nickel
- Selenium
- Vanadium

Calcium is very important for teeth and bones. It is stored in the bones from infancy, peaks in adolescence to middle age then decreases at menopause. The depletion of calcium in the bones decreases their density and bones may become brittle and break. This condition is known as **osteoporosis.** Calcium is found in milk, yoghurt, cheese, nuts, fish and soya beans.

Phosphorous is needed to help build healthy bones and teeth. It also combines with B vitamins to release energy from foods. This mineral is found in eggs, cheese, meat and milk.

Iron is required by the body to combine with copper and protein to make haemoglobin which is used to transport oxygen from the lungs to the body tissues. An iron deficiency is known as **anaemia**.

Functions of minerals

- Form a part of many hormones and enzymes
- Essential for health and growth
- Maintenance of teeth and bones

Activity 3.12: Minerals

1. Why is fluoride added to the water supply?
2. Which mineral is important for thyroid functioning?

RECOMMENDED DAILY INTAKE OF NUTRIENTS

The **recommended dietary intake (RDI)** of macronutrients and micronutrients needed vary with a person's age and lifestyle. The National Health and Medical Research Council have determined a recommended RDI which is considered safe and sufficient for the nutritional needs of people to maintain good health and well-being.

During a person's lifespan nutritional requirements will vary. In times of rapid growth during childhood and the teenage years there is an increased need for foods supplying protein. If a person has a sedentary job there will be a decreased need for energy-rich foods. Conversely if a person has a job where a lot of energy is used or a person plays a lot of sport, there will be a need for extra carbohydrates to provide energy. The needs of a high-performing athlete compared to an elderly person living a low-activity lifestyle are very different.

Pregnancy, health and stress levels also affect the amounts of nutrients needed daily.

What is considered a serving size of various foods?

Milk

1 serve = 250 ml milk

1 serve = 200 gms yoghurt, 40 gms cheese

Meat

1 serve = 65-100 gms lean meat

1 serve = 65-100 gms fish

1 serve = 2 small eggs

1 serve = ½ cup cooked lentils

1 serve = ⅓ cup walnuts, almonds

Vegetables

1 serve = 1 medium potato

1 serve = ½ cup cabbage, broccoli or spinach

1 serve = 1 medium corn on the cob

1 serve = 1 cup other vegetables

Fruit

1 serve = 1 piece of medium size fruit, eg. apple, banana

1 serve = ½ cup 100% fruit juice

1 serve = ½ Tablespoons sultanas

Cereal

1 serve = 2 slices bread

1 serve = 1 cup breakfast cereal, cooked rice, pasta or noodles

1 serve = 1 English muffin

1 serve = 1 crumpet

Discretionary foods

Foods in this category are called empty kilojoule foods as their nutritional content is very low and they are generally high in fat, sugar and salt. Examples are cakes, biscuits, soft drinks, potato crisps, lollies, pastries and alcohol. It is recommended that these foods are consumed only sometimes and in small amounts.

Nutritional requirements of adolescents

Adolescents (12-18 years of age) need a balance of a variety of nutritious foods and plenty of activity to ensure healthy growth and weight gain. It is a period of rapid growth and increased physical activity, therefore there is a greater need for energy and protein foods. It can also be a time of iron deficiency and anaemia, particularly in girls due to the commencement of menstruation and in some cases poor diets. Adolescents also are influenced by peers as to what they eat and this can result in an increased consumption of discretionary food/junk food rather than the more nutritious foods.

Adolescents recommended daily intake

- **Protein g/day**
  - Girls: 45 g
  - Boys: 65 g

Protein content of food
- 1 egg = 13 g
- 1 chicken breast 100 gms = 24 g
- 100 ml milk = 3.3 g
- 100 gms beef = 26 g

- **Calcium mg/day**
  - Girls: 1300 mg
  - Boys: 1300 mg

Calcium content of food
- 250 ml milk = 304 mg
- 250 ml calcium fortified milk = 520 mg
- 1 slice (21 gms) cheddar cheese = 160 mg
- 200 gm low fat yoghurt = 488 mg
- 200 gm regular natural yoghurt = 386 mg
- 90 gms pink salmon in can = 279 mg

- **Iron mg/day**
  - Girls: 15 mg
  - Boys: 11 mg

Iron content of food
- 107 gms chicken = 2.0 mg
- 90 gms tinned salmon = 2.1 mg
- 170 gms beef steak = 5.8 mg
- 110 gms lamb steak = 4.0 mg
- 1 cup broccoli – 0.9 mg
- ½ cup cooked kidney beans = 2.0 mg
- ½ cup Baked Beans = 1.6 mg
- 1 cup cooked spinach = 2.5 mg
- 1 boiled egg = 0.9 mg

WATER

The body is made up of 60% water and is essential to life, even more important than food. It is a major constituent of body fluids, helps maintain the body temperature, is essential for digestion and removal of waste, moistens the air we breathe in, and acts as a lubricant around the eyes.

It is recommended people consume six to eight glasses of fluid a day. As a person's level of activity increases so does the need for extra fluid intake. A person's water intake should also be increased in hot weather and during strenuous exercise to prevent dehydration. Elderly and young children need to be monitored in hot weather to ensure they have enough fluids to prevent dehydration.

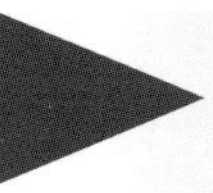

Activity 3.13: Water

1. Water is obtained from beverages (avoid high sugar varieties) and food. What foods have a high water content?
2. Why is the intake of water important?
3. What are the consequences of an insufficient water intake?

FOOD MODELS

Good nutrition is important to maintain a healthy body. To assist people to make healthy food choices, guidelines and food models have been designed for people to follow. They have been developed to emphasise the importance of good nutrition and maintaining a healthy body. Some examples of the more well-known models are outlined below.

A. Australian Dietary Guidelines

The National Health and Medical Research Council (NHMRC) have devised the Australian Dietary Guidelines to highlight groups of foods and lifestyle patterns designed to promote good health and nutrition.

Activity 3.14: Australian Dietary Guidelines

1. What does the Australian Dietary Guidelines promote?
2. Look at the three different sets of dietary guidelines and discuss the differences for the three different age groupings.

B. Australian Guide to Healthy Eating

This model was developed by the Children's Health Development Foundation and Deakin University for the Commonwealth Department of Health and Family Services.

The aim of this model is to encourage the consumption of food from each of the five food groups using the Australian Dietary Guidelines as a reference.

The food groups are:

Food group	
1. Bread, cereals, rice, pasta, noodles 2. Vegetables, legumes 3. Fruit	These 3 groups should provide 80% of the daily food intake.
4. Milk, yoghurt, cheese 5. Meat, fish, poultry, eggs, nuts, legumes	These 2 groups should provide the remaining 20% of the daily food intake.

Remember to drink plenty of water and consume the 'discretionary foods' in moderation.

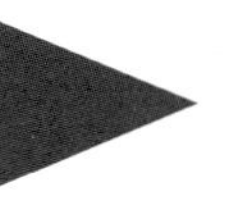

Activity 3.15: Australian Guide to Eating Healthy

1. Draw a circle dividing it into the relevant sections representing each of the food groups in correct proportion.
2. What is the nutritional message being promoted with this food model?

C. Healthy Living Pyramid

This was developed by the Australian Nutrition Foundation (ANF) as a guide to prevent over consumption of many foods and under consumption of other foods.

The Pyramid shows the amounts of food that Australians should consume each day to promote good health. Foods at the bottom of the pyramid should be eaten in the largest quantities each day, the middle in moderate quantities and the top third should only be eaten in small amounts.

Activity 3.16: Healthy Living Pyramid

1. Draw and complete a diagram like the one shown here, detailing what foods are found in each section.
2. What nutrients are found in the 'Eat Most' section?
3. What nutrients are found in the 'Eat Moderately' section?
4. What nutrients are found in the 'Eat Least' section?

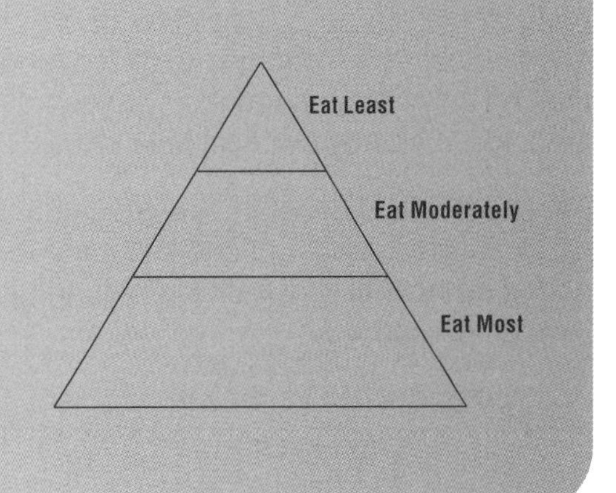

NUTRITION-RELATED HEALTH PROBLEMS

Many present and future health problems are related to food habits. It is important that healthy eating habits are established early in life and followed through the life cycle. A combination of healthy eating and drinking, together with regular exercise, is important to prevent many nutrition related diseases which are related to high fat, sugar, salt and alcohol intake.

What may an individual do to reduce the risk of nutrition-related illnesses?

The most important factor is for a person to be knowledgeable about good nutrition and make wise decisions in the selection of food products.

For example, people with heart disease are recommended to reduce the amount of fat in their diet and increase the amount of fibre in their diet. People with high blood pressure are encouraged to eat a low salt (sodium) diet. High sugar intake has been linked to dental caries and diabetes. In both of these situations people should be aware of the 'hidden' sugars and salt in food. Many processed meats are high in salt and some fruit juices are high in added sugar.

People are advised to consume a diet rich in calcium to guard against osteoporosis. Anaemia is a condition caused by a low iron intake and this is more prevalent in females. To avoid this, red meat and dark green leafy vegetables should be included in the diet.

Labels should be read carefully before purchasing food and a person should use their nutritional knowledge to make wise food choices. Fresh food with little or no processing is generally a better choice nutritionally than highly processed food. Meals should be prepared that are varied, interesting, appetising and nutritious. This will help encourage people to develop healthy eating patterns.

Some of the nutrition-related diseases are described below.

1. Anaemia

Anaemia is due to a deficiency of iron in the diet. When this occurs there is reduced haemoglobin (iron compound in red blood cells) in the body. This carries the oxygen to the body and if this is reduced, oxygen does not reach the body tissues so a person feels tired, lacks energy and has pale skin.

People at risk of anaemia are vegetarians, the malnourished, pregnant and lactating women, females during menstrual years, children during rapid growth periods and athletes.

Foods that are rich in iron are red meat, liver, fish, chicken, wholemeal bread and cereals, dark green leafy vegetables and dried fruits.

2. Dental caries

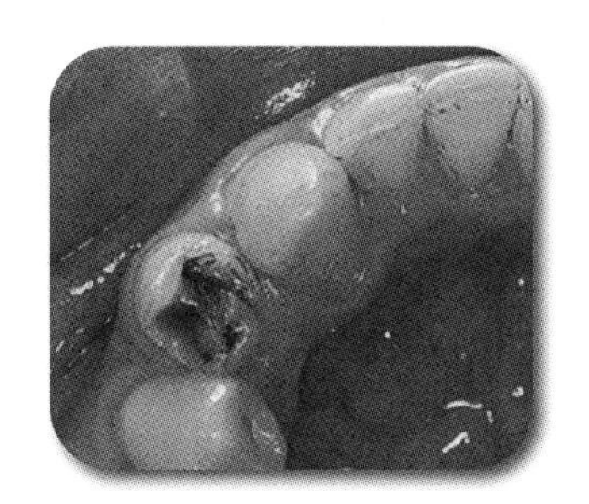

Tooth decay or dental caries is caused by consuming foods or drinks high in sugar which form a film on the teeth. This film causes plaque and encourages the growth of bacteria. The bacteria break down the sugars and an acid is produced which then attacks the tooth enamel causing dental caries.

Dentists recommend that after eating high sugar foods and drinks that teeth are cleaned. It is also recommended that high sugar foods and drinks are eaten all at once rather than over a period of time which prolongs the sugary film on the teeth.

Snack foods are often high in sugar and sticky, causing plaque to form on the teeth.

Activity 3.17: Snack foods

1. List high sugar snack foods and drinks to avoid.
2. Look at the labels on food and list some of the sweeteners and forms of sugars added to the foods, for example honey, glucose.
3. Research breakfast cereals and compare the sugar content in them.
4. Repeat this exercise using cereal bars.
5. Were you surprised with your findings? Make a comment about your results of both the breakfast cereals and cereal bars' sugar content. What advice would you offer from your findings?

3. Diabetes

The body needs a sugar called glucose to provide it with energy and fuel. The body makes glucose from foods containing carbohydrates.

The hormone insulin (produced in the pancreas) helps to transport the glucose from the blood stream into body tissues which actually do the work; for example, brain cells to think, the heart muscle to pump the blood. When insufficient insulin is produced glucose builds up in the blood and when this sugar is not absorbed, a person is diagonised with diabetes.

If diabetes is not treated and managed properly it can lead to blindness, kidney failure, increased risk of cardiovascular disease and limb amputation.

Types of diabetes

- **Type I:** this occurs at any age and occurs when the pancreas cannot produce enough insulin as the cells that actually make the insulin have been destroyed by the body's immune system. People with Type I diabetes need to have insulin injections every day or wear a special insulin pump that releases insulin into the body as needed. It is not a result of poor diet options and shouldn't be linked to Type 2 diabetes.
- **Type 2:** this is non-insulin dependent and is the more common form of diabetes. Type 2 diabetes is a lifestyle disease and is usually associated with high blood pressure and abnormal blood fats. In this situation the pancreas is still making insulin but it is not as efficient in keeping up with the body's demand. This type of diabetes is the more preventable type related to poor dietary habits and lack of exercise.

1. What is hyperglycaemia?
2. Complete a checklist of six points to control and prevent the onset of Type 2 diabetes.
3. Why is the occurrence of Type 2 diabetes increasing in older people?

4. Malnutrition

This is defined as an insufficient intake of nutritious food for energy, health and growth. People most at risk are the elderly, low socio-economic groups and Indigenous people. Malnutrition also occurs when people deliberately restrict their food intake.

Why do you think malnutrition can occur in the elderly?

5. Anorexia nervosa

People suffering from this disease restrict themselves in the amount and type of food they eat. Some people reach the point of starvation and it is most prevalent in teenage girls.

6. Bulimia nervosa

This is the condition where people eat excessive amounts of food and then force the act of vomiting or where a person uses laxatives to prevent weight gain.

Bulimics increase their risk of tooth decay from the vomiting, while other side effects are dizziness and weakness which may lead to heart failure. Bowel problems may also occur from the excessive use of laxatives.

Activity 3.20: Eating disorders

1. Why do you think anorexia nervosa is most prevalent in teenage girls?
2. What are the symptoms of anorexia nervosa?
3. Discuss what you would do if you thought your friend was developing this condition.
4. What are the consequences of a restricted food intake?

7. Overweight and obesity

It is important that the amount of food you eat is balanced with the amount of energy your body is using. If you eat more kilojoules than your body is using you will gain weight. If you are eating less than your body needs you will lose weight.

If you are 3 kg overweight it is like carrying three 1 litre water bottles around. Try it! Nine kgs overweight is equivalent to carrying a car tyre.

Obesity is often a problem when there is a consumption of high fat and/or high sugar foods. The recommendation is that high fat/sugar foods are not included in everyday eating patterns and instead are used only in moderation and infrequently. Also exercise levels need to be sufficient.

Overweight people may face increased risk of the following health problems:

1. High blood pressure
2. Diabetes
3. Cancer
4. Varicose veins
5. Arthritis of the hip, knee and lower spine
6. Heart disease.

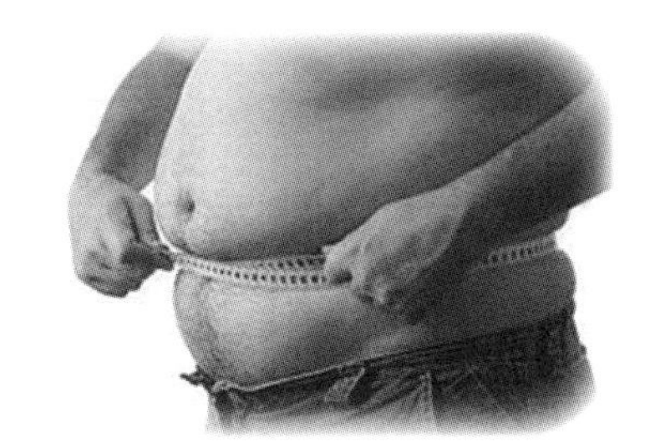

Activity 3.21: Overweight and obesity

1. Write down examples of high fat high sugar foods, for example doughnuts.
2. What foods should be substituted to avoid gaining weight?
3. In today's society much of the weight gain is caused by choosing high energy takeaway foods. Suggest suitable healthy alternatives.
4. What can the general community do to encourage healthy eating and more physical activity?
5. What could be done in the school community to encourage a healthier lifestyle?
6. What cooking methods are the more effective to use to lose weight?

8. Lactose intolerance

Lactose intolerance is when a person has difficulty or the inability to digest lactose – the sugar found in milk and to a lesser degree dairy products.

The enzyme **lactase** helps the body digest **lactose**. Lactase splits lactose into two smaller sugars: **glucose** and **galactose**. These are then absorbed by the body to supply energy. When there is insufficient lactase to break down the lactose, people are said to have **lactase maldigestion**. Undigested lactose passes through the small intestine to the colon where acids and gas form. This causes the symptoms of lactose intolerance which may include bloating, diarrhoea and abdominal pain

People with lactose intolerance do not need to eliminate all dairy foods from their diets. Cheeses contain very little lactose and yoghurt is well digested as the bacteria ferment the lactose. Research has shown that people with lactose maldigestion can drink up to two cups of milk a day without experiencing any symptoms.

Milk, cheese and yoghurt play an important role in a balanced diet as these foods provide protein, minerals (calcium, zinc, phosphorous, magnesium and potassium) and vitamins (A,B12 and riboflavin). Removing dairy foods from the diet can lead to an increased risk of osteoporosis. The amount of lactose that can be tolerated varies from person to person.

People who are lactose intolerant should drink milk with other foods rather than on an empty stomach, have small serves of milk during the day, and start small and gradually build up milk consumption to build up their tolerance. Regular fat milk is better tolerated than skim or low fat, and yoghurt and cheese are often better tolerated.

9. Coeliac disease

Coeliac disease is when the immune system reacts abnormally to gluten and causes damage to the small bowel. **Gluten** is the protein found in wheat (including spelt), rye, oats and barley. It can affect people of all ages and both males and females. About 1 in 70 people in Australia have coelaic disease. There is an increase in the number of people with this disease due to a true increase in the incidence of coeliac disease and better diagnosis rates.

Once diagnosed, people remain sensitive to gluten throughout their lives. The condition can be successfully managed with a strict gluten-free diet.

Fresh fruit and vegetables, fresh meats, eggs, nuts, legumes, fats and oils, milk and gluten-free grains such as corn and rice are gluten free foods. Those with a gluten intolerance should always purchase foods that are labelled gluten free, avoid labels that state 'may contain gluten' and always check the ingredient list before purchasing food.

Vegetarian diets

Vegetarian diets relate to peoples' various attitudes and eating behaviours with regard to food of animal origin. All vegetarians base their diet on foods of plant origin.

There are five major categories of vegetarian diets.

1. **Semi-vegetarian:** eats no red meat but eats poultry, fish, dairy foods and eggs.
2. **Lacto-vegetarian:** eats dairy foods but no fish, eggs, red meat or poultry.
3. **Lacto-ova vegetarian:** eats eggs and dairy foods but no fish, poultry or red meat.
4. **Pescetarian:** eats fish and other seafood but no meat or poultry. They may or may not include dairy foods and eggs.
5. **Vegan:** eats foods of a plant origin only.

People may choose to become vegetarian because of religious beliefs or because they believe meat is harmful to health, especially heart health. A moral reason is that animals have the same right to live as humans. Lacto and Lacto-ova vegetarians consider themselves more humane than meat eaters as dairy foods and eggs can be used without killing an animal. Vegans will likely follow the moral argument but also choose a diet low in saturated fat which will decrease the amount of cholesterol in the diet. Plant foods contain little saturated fat (coconut oil and palm oil are exceptions) and no cholesterol.

Vegetarians believe the higher level of vitamins, anti-oxidants and fibre found in plant foods is nutritionally beneficial and there is reduced environmental degradation with a vegetarian diet.

It should be noted vegetarian diets can be low in iron which may lead to anaemia. Vegans may also have a diet low in calcium, zinc, Vitamin B12 and protein.

People also choose to be vegetarian as it is cheaper than a diet with animal based foods.

What foods should be included in a vegetarian diet to provide all the required nutrients?

FACTORS AFFECTING NUTRITION RELATED HEALTH ISSUES

1. Lifestyle

Many people these days are 'time poor' due to employment and changing lifestyle. The foods we eat are also influenced by household structures;the roles of family members, interests and education. In today's society people do not have the same amount of time as in previous years for meal preparation and many factors influence what we eat.

People are away from home at mealtimes and they may be limited by the food available, for example food sold at work, sporting events, concerts and school canteens.

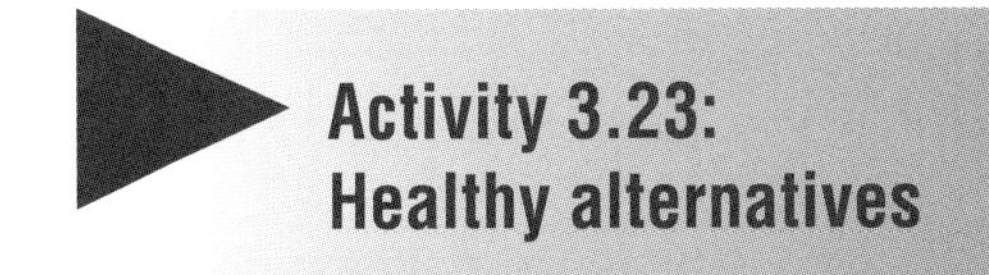

What are some of the healthier alternatives available at sporting venues in regard to food and drinks?

Cheap restaurants and cafés have become a substitute for meal preparation in the home. Dining out has become a way of life. As a consequence, the number of cheaper cafés and restaurants has increased over the last few years. There is also an increase in the number of takeaway meals being sold.

Supermarkets have also adapted to the change in lifestyle by increasing the amount of partially or fully prepared meals available.

Activity 3.24: Prepared meals

1. Look in your local supermarket and list some of the partially or fully prepared meals that are available.
2. List some of the advantages and disadvantages of this trend of less meal preparation being completed in the home.

2. Advertising

Advertising and peer pressure also influence teenagers' food choices and many snack type foods become fad foods due to this. Often advertising focuses on low nutritional food and people spend their money on these items rather than the more nutritious food which is invariably cheaper when you compare price per kilo.

Supermarkets are also very clever at encouraging the consumer to buy foods they may not really need.

Advertising marketing strategies are discussed in more detail in other sections of this book.

Activity 3.25: Advertising

1. What foods have you been influenced into eating by advertising and friends?
2. Take a trip to your local supermarket and analyse the selling tricks they use. Compare your findings with other class members and compile a list, for example impulse lines by a checkout.
3. Using a table, list the healthy choices from each of the various sections in a supermarket – Meat, Bakery, Fruit and vegetables, Snack foods, Delicatessen, Chilled and frozen food, Dairy.

3. Location

The area where a person lives determines food availability. Rural communities often pay higher prices due to transport costs from city markets and the quality may also be inferior due to storage requirements, particularly for perishable foods. Some foods may be too expensive to be purchased regularly; for example, lettuces are expensive on Christmas Island. In smaller communities it is not economically viable to store a wide variety of food, especially perishables, in the local supermarket. Think of small places such as Babakin, Kalumburu, Cuballing and Lake King.

Some places may be inaccessible at certain times of the year due to weather conditions, which would limit the variety of food available, for example, communities along Gibb River Road.

Activity 3.26: Food availability

1. What are some of the foods which would not be as readily available in a smaller country town compared to larger centres?
2. Over the years with improved technology, food availability (particularly perishables) has become increasingly consistent all year around. What are the changes that have made this possible?

4. Culture and religion

Foods form a large aspect of most cultures and/or religious beliefs.

Culture

In some cultures it is desirable to be super thin and this leads to health problems such as bulimia and anorexia whereas in other countries being overweight is a sign of affluence. Culture can also influence the type of food eaten and methods of cooking used. This is discussed in more detail later in this book.

Religion

Many special foods are consumed to celebrate special religious occasions. Some religions also ban some foods from being consumed all the time or for certain times of the year.

5. Economic factors

A person's occupation and finances influence the type of food bought. It should be noted that wise food choices do not always mean spending large amounts of money. Careful planning and nutritional knowledge is essential to maximise healthy eating.

The level of affluence also influences the type of food preparation and processing equipment available. The more energy-dense foods are usually the cheapest, for example crisps and biscuits, and this often results in people satisfying their hunger with these foods which may lead to overweight issues.

MODIFYING EATING HABITS

It is not always easy to change a person's eating habits and to make the change successful a person must be motivated, persistent and be prepared to deal with the occasional failure.

Choose goals which are achievable and include food that you like but served in a healthier manner. Success is more likely with smaller sequential goals rather than a large goal. Once one goal has been achieved and maintained then move on to the next one. If you have a lapse in achieving your goals remember the benefits to changing your eating habits. Choose strategies that suit your particular lifestyle and taste. Remember you must take ownership of your eating habits and instigate changes yourself.

Activity 3.27: Eating habits

1. Analyse your eating habits and decide on any changes you should make to your diet to improve your long-term health. Determine how you will go about making the necessary changes. Write out your goals then put your plan into action. After a period of six weeks analyse how you have gone. At the end of each week write done how you feel you have gone and note if there has been any lapses. Why did these happen and how did you deal with them? [HINT: use a table to record and summarise your findings.]
2. How did you go?
3. What were any hindrances to your success?

EXTENSION

Extension Activities

1. Develop a **healthy food snack** that will keep for a couple of days. Justify your choice. Make it, and then do a sensory evaluation on the day of producing then also two days after. Ask other class members to also taste your product. Reflect on your product as to its suitability for a teenage snack and its keeping qualities. Include your recommended storage choice. Include the recipe and refection.
2. Interview your parents and grandparents to determine how their **food habits have changed over the years** and the reasons for the changes. Also determine how their cooking skills have changed over the years. Discuss the nutritional status of their meals previously compared to present times.
3. Your father has been diagnosed with **high blood pressure** and your mother's family has a history of **heart disease.** Using your nutritional knowledge discuss foods and cooking methods that they should avoid. Also suggest some healthy food choices and lifestyle changes they should incorporate into their daily life. Give some examples of a healthy daily eating pattern.

CHAPTER 4
The Technology Process

Key Concepts

- **The Technology Process**
- **Food production and planning**
  - **meal planning and the design process**
  - **today's meal planning issues**
  - **considerations and constraints in meal planning**
- **Investigating raw and processed food products**

THE TECHNOLOGY PROCESS

The ever-evolving face of food is a result of continually meeting the changing needs and requirements of today's consumers. This is often dealt with, even on a subconscious level, with the **Technology Process** which, through thought and application, systematically processes options and considerations to trial results and evaluate their effectiveness.

Food technology refers to the use of equipment, skills and food to improve, maintain or alter products to meet the consumer's needs.

The **Technology Process** is defined as the steps taken for solving problems and involves the development of ideas and the creation of solutions.

The Technology Process is made up of four sections:

1. **Investigating:** this is where, once given a problem, an assessment of the nature and circumstances surrounding the problem is undertaken. This is often in the form of a design brief. Research is completed whereby information that is helpful in addressing the problem is gathered and analysed. Once all relevant information is available and background knowledge is attained, the next process is started.
2. **Devising:** this is the design stage (or design process) of the Technology Process whereby plans are developed that solve the problem or issue. Because problems rarely have one solution, options are considered factoring in constraints and/or priorities, and predictions are made about possible outcomes. Upon the selection of one possible solution to the problem resources are identified and selected for use in the production phase, and evaluative criteria are also identified at this stage.
3. **Producing:** taking the developed plan from the devising stage, a product or process is created using resources identified as well as skills and knowledge of the participants.
4. **Evaluating:** this involves using the criteria developed from the devising stage to determine whether the solution developed meets all of the needs of the problem. It is at the evaluating stage that the Technology Process may be repeated by starting again in at any of the above three stages.

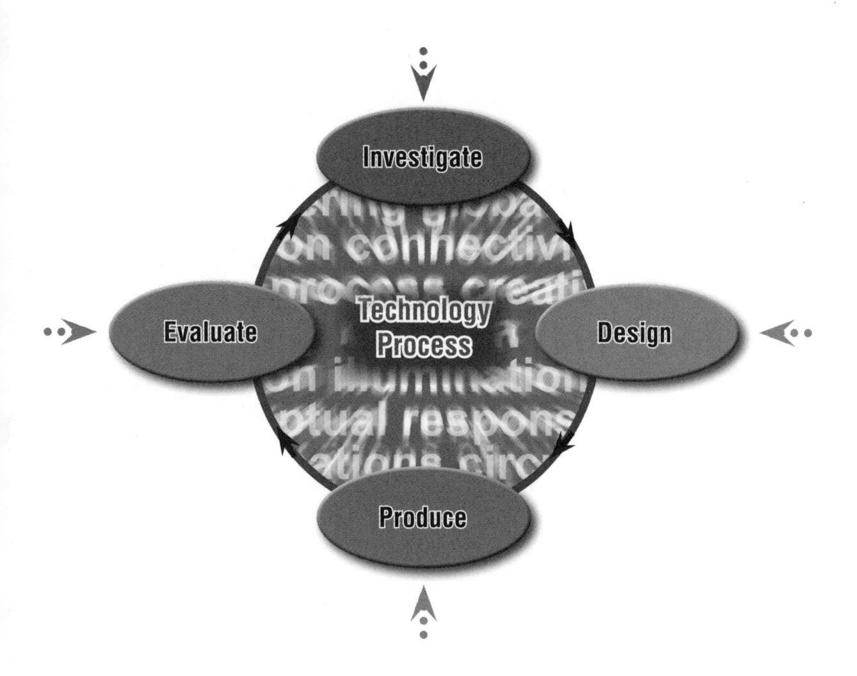

To make this simpler, look at the process and its application as set out here.

The Technology Process is **cyclic** because you can enter the process at any stage. Hence solutions are continually evolving and can be related to the production of new and/or improved food products by food manufacturers, for example, the changing nature of bread products produced from local manufacturers. They investigate consumer trends and needs; create a saleable product; test it out on consumers; and modify it according to the consumers' response.

As per the diagram, the external arrows indicate that entry to the Technology Process can be at any of the four stages. The process continues when you refine and improve results, so the diagram is called **cyclic**.

Activity 4.1: Using the Technology Process

1. Go to your local bakery and look for their new products. They are often offered as free sampling. Select one and see if you can work out why they are trialling it as a new product. Explain why or why not you think it is going to be successful and who you think the product is being marketed to.
2. Follow the process by completing the application in the boxes provided. 'Producing' can be done at home.

	Description	Application
Investigating	(Assessment of the nature and circumstances of the problem or need.) • What is the problem or need? • What information do I need to collect and analyse? • Where can I get my information from? • What factors do I need to consider?	I am on duty to provide tea tonight for my family...

	Description	Application
Devising	(Generation of ideas/plans/proposals for solving the problem.) • Can I come up with an idea or a plan/proposal? • What priorities and constraints do I have? • What alternatives do I have? • Which is the best one to achieve the result I need? • What consequences could this have?	
Producing	(Translation of designs and plans into products and processes.) • Produce the item and test it. • Apply techniques, skills and knowledge. • Use adaptive techniques as required.	
Evaluating	(Application of criteria to measure and test products and processes.) • Is the end result suited for the purpose? Is it appropriate? • How can I make the product better? • How can I do this more efficiently and effectively? • What parts of the process were the most successful? • How can I improve this next time? What are some suggestions?	

Review – The end result may need modifications and/or improvements. You may need to repeat the Technology Process to further refine your product.

3. Devise several pointers that you would use to assess the effectiveness of your final decision made above.

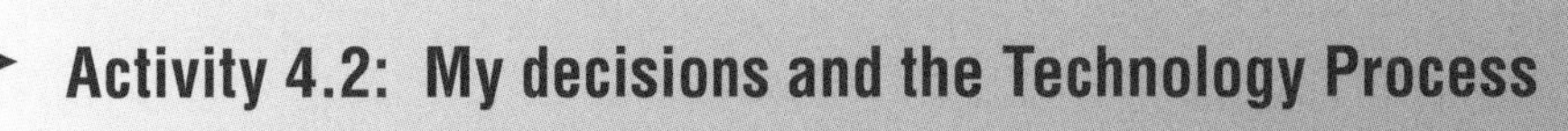

Activity 4.2: My decisions and the Technology Process

In the activities set out below, write anecdotal notes rather than paragraphs to indicate your thinking and actions.

1. Think back to the last time you prepared an afternoon snack. Describe your thought processes as to how you made a decision about what to eat and what influenced your thinking. Think about your values and attitudes to food. Now write this into the Technology Process with an evaluation of your end result. (Detail your response as set out below.)
   - Investigating (Gather information.):
   - Devising (What are the options?):
   - Producing (What do you have to do to test an option?):
   - Evaluating (Was it successful or how could you do it better?):

2. Your friend's birthday is in three days and you've been asked to provide the birthday cake at school. Your friend, however, is allergic to wheat. Using the technology process work through your options so that you come to a suitable solution.
   - Investigating (Gather information.):
   - Devising (What are the options?):
   - Producing (What do you have to do to test an option?):
   - Evaluating (Was it successful or how could you do it better?):

3. Your local café is trying to entice the teenage market to its store and is offering a competition with food vouchers for a year for the winner. You need to come up with an enticing lunch suitable to their needs and values. With a group use the technology process to come up with ideas and test one out. Present it to the class and obtain their feedback. What adjustments, if any, do you have to make?
   - Investigating (Gather information.):
   - Devising (What are the options?):
   - Producing (What do you have to do to test an option?):
   - Evaluating (Was it successful or how could you do it better?):

4. You have started a new swimming program that requires you to train early in the morning. You need food for energy but find breakfast sometimes too hard or you often don't have time to eat it. Your coach is encouraging you to be healthy and include breakfast every day. Work out solutions to your dilemma and come up with a plan. Go through the process below.
   - Investigating (Gather information.):
   - Devising (What are the options?):
   - Producing (What do you have to do to test an option?):
   - Evaluating (Was it successful or how could you do it better?):

FOOD PRODUCTION AND PLANNING

Food and recipe management are key elements to a successfully run kitchen. They are also key elements of the Food Technology Process. In any form of food preparation and management aspects such as time, energy and money are simply some of the resources that can be saved through efficient and effective running of your kitchen. Concepts that are applied in large kitchens or food production areas can also be applied to the domestic kitchen, for example, from the ordering of stock (buying of ingredients) through to staff dynamics (who is going to cook).

Let's start at the basics.

Meal planning and the design process

This is a process that is applied to all food production – whether it is a conscious process involving a series of complex decisions or a simplified reaction to a situation. From planning and managing a function serving 500 guests to creating a simple snack for one, the planning process is intrinsic and is a reaction to a problem that requires solving. This is a basic form of the technology process in operation.

A **Design Brief** – this is usually a statement or description about a problem that requires solving and incorporates the purpose of the food preparation and any considerations or constraints that need to be applied.

Design Process is another name for the Technology Process whereby a person is given a problem, identifies possible solutions, and trials an idea by producing a solution and evaluating the result. This then may be modified for future trials or use.

Today's meal planning issues

Compared to generations ago today's family is one that generally follows a simplified meal planning process. With the current family lifestyle being one that incorporates busy individual schedules, meal planning has reflected this with increased demand for takeaway meals, out of home dining and convenience foods. Even though it makes sense to plan meals for the coming week or fortnight, many people don't have this luxury and tend to make last minute decisions that may end up costing more than if the meal was planned beforehand.

Activity 4.3: Meal planning

1. Compare your meals eaten over a few weeks, during which for some time you do not plan meals and live day-to-day, and during other periods of time where you can plan at least three days' meals. What do you notice about the effect both styles of decision making have on resources such as money, time and meals as well as on the quality of the food you are eating? Make conclusions about what you have found out. You may need to apply this to your lunchtime eating habits.

2. Talk to a grandparent or a person from around that generation and ask them how meal planning is different from their teen years to that of today. What aspects were better then and what is better now? Maybe get them to describe a typical day's meal for them as a child. How is this different from yours?

Meal planning in today's society has changed considerably over time. Australians are developing a more sophisticated food culture whereby they are being exposed to more cultural influences; they have a wider range of foods available for purchase; restaurant dining is being scaled down to two or three courses; eating out is providing many options such as tapas menus and raw food restaurants; and we are being influenced by lifestyle and media in choices for our snacks and meals.

Takeaway meals are those that are prepared in a commercial kitchen with or without a restaurant attached. Patrons have the choice of taking the meal home to eat.

Convenience foods are foods that have had some partial preparation completed for you. These can include vegetables that are washed and sliced for you through to frozen meals that simply require heating before serving.

Resources are elements that are used to assist in the solving of the problem. They are human or non-human.

Considerations and constraints in meal planning

In making choices and decisions when meal planning, there are general factors that need to be incorporated into the design process. These factors influence each other and therefore the choices that will be made. For example, if you have a lot of time to prepare a meal you may decide to produce your own foods that could otherwise be partially prepared for you such as pasta, custard, sauces or pastry.

Rarely does a home meal contain everything that 'comes from scratch', rather we opt to use partially processed food products to save us time and, also, simply for the convenience.

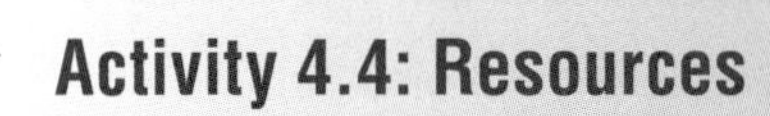

Activity 4.4: Resources

1. Research the term 'Paleo Foods'. What does it mean? Describe the benefits from consuming a Paleolithic Diet. Generate a list of Paleo foods, and see if you can generate a day's menu using these foods only.

2. Define and explain each of the resources below as to how they contribute to the design process when in food preparation situations.
   - Time
   - Resources
   - Ingredients
   - Skills
   - Knowledge
   - Equipment
   - Money
   - Other

3. As a young adult, explain how the resources listed below affect your food decisions for mealtimes.
   - Resources
   - Time
   - Ingredients
   - Skills
   - Knowledge
   - Equipment
   - Money
   - Other

Constraints

These are described as factors that limit your choices or direct your planning in certain directions. For example, a constraint to planning a meal for a friend could be that they are a vegan, or that they may be an athlete on a modified diet, or that they work in the evenings whereby lunch is the only alternative that you can do for them.

Examples of constraints are:

- time of year – foods available, ie. seasonal
- climate – what can grow in certain climates
- who the food is for – age
- dietary requirements/health concerns/energy needs
- how many persons
- food elements
  - colours of dishes together
  - food flavours, ie. contrasting, complimentary
  - texture – 'crunch' factor needed!
  - overall presentation.

Activity 4.5: Constraints

1. Think about your family. Describe what constraints you would have to consider if you were to cook them breakfast next Sunday morning.

2. In a group of six, you are to plan an end of year function, a dinner dance for the 120 students in your year to be held in the school hall and catered for by your class. Work out what needs to be decided, and who needs to do what. Write a plan for the event and a budget for all the requirements of the evening. Make up a menu and trial your recipes for the rest of the class to sample. Answer the following questions.

   a. What decisions had to be made initially? What were they and how were they decided?
   b. Who became the decision makers (leaders) of the group and who became the workers? What contribution did each group member provide?
   c. What difficulties were experienced and how were they overcome? Did the problem solving process work satisfactorily or could it have been managed more efficiently?
   d. Outline what help (or staff) you would need on the evening and how this could be coordinated.
   e. What would be the key elements required for a successful event?
   f. Imagine on the day of the evening you were suddenly alerted that you had twenty special guests that you had not anticipated. What could be done to solve this issue and how would you overcome the problem of catering for the extra people?
   g. Catering for functions is something that requires a lot of planning and thought. Make up a list of guidelines for others to follow to help them if they ever had to cater for a function themselves.

INVESTIGATING RAW AND PROCESSED FOOD PRODUCTS

An aspect of food processing and preparation is the decision making that goes into the selection of ingredients. Historically, the seasons have dictated what foods were available for use, with fresh fruit and vegetables only used when 'in season'. The quality of certain meats and dairy products also varied seasonally, and egg production was affected by external factors as well. With increasing technological developments, we can now enjoy many foods all year long. But are foods as good for you if they are processed and/or stored? Do they lose nutritional value and quality of texture but are viable because they increase their shelf life? Does the heating of foods deteriorate their value in our diet?

How about the use of pre-prepared foods or convenience foods such as minced garlic, pesto sauces and pastry to name a few? Are they a viable option to use or do they affect the final quality of the dish prepared?

Let's look at the food options available.

- **Raw foods:** these are foods that occur in their natural state, typically being fruit, vegetables, meats, eggs and legumes. These foods are those that are typically picked or harvested and are used fresh, or in their natural state with minimal processing. Chicken pieces are considered a raw food even though they have been prepared ready to use.
- **Processed foods:** these are foods that have been altered in a way to increase their value to the end user, such as storage, appearance or palatability. Processing can be as simple as washing (eg. potatoes) or a much more involved process that creates an end product such as a frozen instant meal. Another term for these foods is convenience foods. Most foods that are store-bought have been semi-processed or processed in some way, if not simply picked and cleaned, creating a convenience to you – the end user.

A newish ethos is that of the 'Raw Food Diet'. This diet uses foods that are usually organic and vegan in nature and involves processing these foods below 46° C so that their nutritional value remains intact. Raw foods may need some preparation to make them more edible, such as peeling; cutting; or blending. They may also need to be mixed with herbs and spices to make them more palatable. The benefits involve getting the most nutritionally from the available food sources and maintaining a chemical/additive free diet.

What to use?

When deciding upon foods and ingredients when planning food items, consideration needs to be given to:

- nutrition
- cost
- time
- use
- shelf-life of the ingredients.

Ideally, to create an awesomely nutritional and appetising meal, the ingredients would be available from local suppliers, fresh that day, and you would have a lot of time available to prepare them correctly. Local fresh food markets are obviously a great food source for obtaining these natural, raw and fresh ingredients. However, if there is an ingredient that you cannot source from these markets, you may need to look at alternatives that are processed and stored in some way.

Activity 4.6: Raw and processed foods

1. Select a vegetable or fruit, and in a table investigate the different forms it can be bought in, comparing the cost, nutritional information, shelf life and uses. Draw conclusions comparing the raw product to the processed products.

2. Plan a meal that uses raw ingredients only. Source semi-processed and processed foods that could be substituted in the meal and compare cost, time prepared to make and nutritional implications. Which would you prefer and why?

EXTENSION

Extension Activities

1. Some people often find the thought of entertaining others in their own home quite daunting, especially if it involves cooking for a special event. With practice it can become easier and less intimidating. To help yourself gain practice at this type of event, plan your next birthday party to be held at home. The variables that you will have to decide upon will be the aspects such as who you will invite, when will it be, what type of party you will want and so on.

   Use the technology process to come up with a set of alternatives and then work your way through the process to come up with a final decision. Complete the planning and carry through with the party, even if it is not your birthday! After all, it is a practice.

2. With our lifestyles getting more and more complicated and families experiencing increasing demands and pressures, food is often a resource that is overlooked and paid scant attention to.

   Explain how your own lifestyle influences your food choices. Think about your activity levels, education, sporting, hobbies, friends and family. What aspects of your diet are healthy and suitable? What aspects are not as healthy and suitable? Remember – food is meant to be nutritional and fuel for the body. What goals do you need to set for yourself nutritionally? Develop a healthy food plan that reflects your lifestyle, goals and food needs. Make the plan for a week to cater for school days as well as weekend events. Remember to include foods from a wide range of sources. Prepare some of your meals and explain how they reflect your food plan. Are they reasonable options for you? Think of time taken to prepare, skills you have, costs of the ingredients, appeal and so on. Do you need to improve your healthy food plan or adapt it to further suit your needs? Imagine that you need to increase iron in your diet. How would you do this?

3. A lot of parents are sending their primary-aged children to school without swufficient breakfast or simply none at all. The principal of your local primary school has asked your nutrition teacher for some help in informing his parent community about the importance of breakfast.

   Use the Technology Process to develop a plan on the best way to inform his parent body on the importance of breakfast and also develop some school plans to reinforce this aspect in their curriculum.

Useful Websites

http://therawfoodinstituteofaustralia.com/category/raw-recipes/: Outlines the raw food concept plus includes recipes and ideas.

www.healthyfoodhealthyplanet.org/: Contains a meal planning tool that assists in weekly planning incorporating dietary needs and healthy food options

www.health.sa.gov.au: Contains a menu template, sample menus and recipes

CHAPTER 5
Devising Food Products

Key Concepts

- Defining the preparation techniques
- Terminology
- The recipe
  - adjusting quantities
- Food ordering
- Costing
- Weights and measures of common ingredients

DEFINING THE PREPARATION TECHNIQUES

When you encounter a recipe it is often filled with terminology that requires careful attention. This preparation terminology states what it is that you have to do with the listed ingredients and equipment. Failure to follow these terms or processes may mean a failure to complete the recipe to the standard expected! Take eggs, for example. If you wanted to make a fluffy or soufflé omelette, if you **mixed** the eggs instead of **beating** them you would end up with a flat French omelette or egg pancake instead of a fluffy omelette: two different products from the same ingredient.

TERMINOLOGY

A lot of our preparation terms are derived from French origins. Because France has a history of classical food preparation techniques, much of the world has adopted the French phrasing and has incorporated it into their own language. For example, mis en place is a French term used extensively in the food industry and relates to the preparation of ingredients before processing them in recipes.

Activity 5.1: Recipe terminology

Look at the recipe given below. Underline the preparation terminology you encounter, that is, the terms that tell you how to prepare the ingredients.

Beef with Green Peppercorn Sauce

1 clove garlic
4 thick fillet steaks
½ c orange liqueur
55 g can green peppercorns
½ c pouring cream

1. Cut garlic in half and rub over the steaks.
2. Melt butter in a frying pan.
3. On a high heat sear the steaks cooking until desired.
4. Pour over the liqueur, ignite and allow the flames to die down.
5. Remove steaks to a platter and keep warm.
6. Stir in peppercorns and a little extra liqueur to deglaze the pan.
7. Stir in the cream, adjust the seasonings if necessary.
8. Reduce heat and stir for 2-3 minutes until the pan juices are blended.
9. Spoon over steaks and serve with steamed vegetables.

Activity 5.2: Cooking terminology

With the help of reference material, explain the terminology below relating it specifically to food preparation.

- Bake
- Barbeque
- Baste
- Blanch
- Blend
- Brown
- Bruise
- Brunoise
- Chop
- Coat
- Cream
- Crumb
- Crush
- Deglaze
- Dice
- Dissolve
- Drain
- Dust
- Flip
- Freeze
- Garnish
- Glaze
- Grease
- Grill
- Jardiniere
- Julienne
- Macedoine
- Marinade
- Mash
- Melt
- Microwave
- Mirepoix
- Mis en place
- Mix
- Parboil
- Peel
- Pierce
- Poach
- Pull
- Refresh
- Refrigerate
- Roast
- Rubbing In
- Saute
- Sear
- Scald
- Score
- Segment
- Serve
- Shred
- Sift
- Simmer
- Slice
- Steep
- Stir
- Toss
- Well
- Whip
- Whisk
- Zest

THE RECIPE

Recipes come in all forms, some of the quaintest being those handwritten on the back of old envelopes or whatever and passed down through the generations! There are so many versions or recipes of dishes that it can become quite confusing as to what works. This will eventually come down to experience and what your preferences are with specific dishes.

A **recipe** is a set of instructions detailing how to cook a particular dish from a list of ingredients.

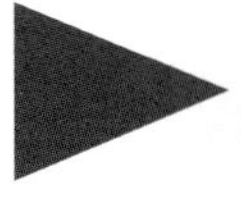

Activity 5.3: Recipes

See how many recipes of a certain dish you can find (eg. sponge, lasagna, Caesar salad) and briefly describe the differences in ingredients and processes. Looking at them try, to determine one that you like and explain why.

Parts of the recipe

- **Name:** usually a description of the dish or food.
- **Servings:** how many people it will feed, or how many portions it will make.
- **Preparation time:** the time taken to prepare the dish.
- **Cooking time:** time the food requires in the oven.
- **Ingredients:** a list of foods required and their quantities in measurement
- **Method:** how the dish is to be prepared.

Some recipes also include extra information such as:

- **Variations:** what different ingredients can be added to produce a slightly different result.
- **Dietary information:** the kilojoules of the servings plus specific information regarding nutritional details such as fat, sugar, salt content, presence of nuts or gluten.
- **Storage information:** how long it will keep.
- **Microwave instructions:** whether the recipe can be adapted for the microwave or not.
- **Equipment needed:** often specialised pieces.
- **Accompaniments:** what it can be served with.

How all of this information is presented varies, however the most easily recognised format is set out below.

Name/Title	Frittata
Servings	Serves 6 to 8
Preparation time	Preparation 10 minutes
Cooking time	Cooking time 50 minutes
Ingredients	8 eggs, beaten lightly 125 ml cream 2 spring onions, thinly sliced 2 roma tomatoes, thinly sliced ¼ cup fresh basil, shredded 180 g feta cheese salt, freshly ground black pepper
Method	1. Preheat oven to moderate. Grease a 20 cm square baking dish. Line with baking paper. 2. Combine the eggs, cream, onions, basil, cheese, salt and pepper in a large bowl. Mix lightly. 3. Pour mixture into prepared dish and top with sliced tomato. 4. Bake in moderate oven for about 50 minutes or until browned and set.
Serving ideas	5. Serve with a green salad.
Storage	Keeps for one day refrigerated. (Not suitable to freeze)
Microwave	(Not suitable to microwave)

Not all recipes are set out like this, however this format appears to be most popular because the information is readily seen and the layout appears logical.

Activity 5.4: Recipe books

1. When you last had to plan a meal and use a recipe, where did it come from? Where do you usually look for your recipes? Why do you choose that source?
2. Think about the recipe books that you may have been using. Describe your favourite book and why this is so (think not only about the foods involved but other factors such as presentation, successfulness of recipes). What makes a recipe book useful and fun to use? Why are there some that you do not like to use?

3. Come up with an innovative way to present recipes in the kitchen. Think in terms of technology, for example splash-backs as interactive displays.
4. Think of how your teacher uses recipes in the classroom. Often it involves photocopying and the page getting messy from the cooking activity. Outline an innovative way that a cooking class can have access to recipes without the need for photocopying. Think of a multimedia solution and outline its benefits over your current system.

ADJUSTING QUANTITIES

A form of mathematics that comes into the kitchen involves that of recipe quantity adjustments. It is not that easy though, and there are a few tricks to this in that sometimes the special flavouring ingredients need to be adjusted carefully to maximise their potential. At times, if you are not sure, it is safer to add too little and adjust upwards rather that adding too much and possibly ruining the dish.

Activity 5.5: Adjusting quantities

1. Brainstorm with the class and come up with how you can 'save' a dish that has had too many herbs and/or spices added. List your ideas below.
2. Complete the recipe quantity adjustment below by filling in the table.

Ingredient	Serves 4	Serves 1	Serves 50	Method
Butter	30 g			1. Prepare vegetables by slicing and chopping into cubes. 2. Melt butter in saucepan and sauté the vegetables and bacon for 5 minutes. 3. Add rinsed beans, stock and tomatoes. 4. Bring to boil, cover and simmer 30 minutes. 5. Add macaroni and cook further 25 minutes gently. 6. Serve in deep bowls sprinkled with chopped parsley and shredded parmesan.
Carrots	2			
Celery sticks	2			
Potatoes	1			
Onions	1			
Garlic cloves	1			
Bacon rashers	1			
Beef stock	1.5 L			

Ingredient	Serves 4	Serves 1	Serves 50	Method
Red kidney beans, canned	½ c			
Peeled tomatoes, tinned	½ c			
Macaroni	$^1/_3$ c			
Parsley	2 Tbsp			
Pepper	¼ tsp			
Salt	¼ tsp			
Parmesan cheese	2 Tbsp			

3. Define the preparation terms used in the above dish:
   - slice
   - chop
   - melt
   - saute
   - boil
   - simmer
   - shred.
4. Which ingredient/s have to be considered very carefully when converting recipes to large amounts and why?
5. What do you think is the product produced by this recipe?

FOOD ORDERING

Cooking is, at times, not as easy as going into the kitchen or food preparation area and deciding there and then to make a dish. Many organisational and resource factors have an impact on whether or not you can do this. Factors such as the time available, equipment on hand and ingredients currently available for use can turn your inspirational cooking mood into an unsatisfactory experience. Therefore, a chef and home cook need to manage their food resources carefully. Too much stock on hand may lead to foods not being used and therefore wasted whilst not enough stock on hand creates supply or food service issues and general frustration.

For the home kitchen, many households manage with one grocery shop per week, with additional items being picked up as required. For those living in remote or isolated country areas, shopping may only occur once every 2-6 weeks, with 'fresh' items either being carefully stored or being produced locally.

Some people choose to shop daily for foods, which is a habit of a lot of Europeans who have access to fresh daily markets of fruit and vegetables, dairy produce, fish and poultry to name a few suppliers. This luxury is not available to everyone in Western Australia.

As part of the food production process, the ordering of food comprises a vital aspect requiring accuracy and thoroughness. When working in a commercial kitchen or that at school, the stock on hand is often purchased as required with limited or no excess available. This helps to keep food stocks turning over with minimal wastage. It also frees up space in a kitchen and assists in the budgeting or financial management of the kitchen operation. Therefore, if food ordering is not accurate and a food worker assumes ingredients are available (items such as flour, oil, dried herbs are most common), a chef may later find the kitchen has run out of the missed ingredient and disasters start to happen. Taking a 'spare' few tomatoes from the fridge may mean another dish misses out when the ingredient may be a focal point and cannot be substituted with something else.

It is important to become familiar with the food ordering system applied by the school or workplace you work in. Remember – if you are not sure that you can use items that may be on hand or in stock, ask questions and don't assume it is like your kitchen at home where it may not matter if you take food items without asking permission first.

For the classroom kitchen, some of the same issues apply as to those for the kitchen at home. Most home economics departments run on an ordering system whereby the food is planned and ordered at least a week before it arrives and is put to use. Therefore, there is minimal extra stock kept in supply, making spontaneous cooking or alterations to menus difficult to cater for.

Activity 5.6: Food supply

1. Consider people living in Newman compared to those living in Esperance. What differences would each community have regarding food supply? Investigate how people in these communities would receive fresh goods such as fruit and vegetables, meat and dairy foods. What food sourcing recommendations would you have for people moving into these communities to live permanently?

2. Find out about the food ordering system at your school. Identify aspects such as what is done manually and electronically, and who does what. Where does the staff source most of their ingredients from and do they use specialty outlets, etc.? Write a description of the whole process from student order to ingredient supply ready for food preparation.

3. From the information above, complete a PMI chart outlining the food ordering system's characteristics.

4. From this information outline several improvements you would recommend the school to implement to make their food management system operate more smoothly.

COSTING

Your teacher will often give you tasks or challenges to prepare food within a budget. Initially the process of costing may seem difficult, however, once you become familiar with the formula below it becomes quite simple. The ability to accurately cost food items is a vital aspect of the food preparation industry as well as your home economics centre.

Costing is a relatively simple process based upon a common formula set out below. Use this to calculate individual ingredient costs.

$$\text{Ingredient cost} = \frac{\text{Original item total cost}}{\text{Quantity purchased}} \times \text{Amount required}$$

(Or simply put as: original item cost divided by quantity or weight purchased times the amount required.)

Note: You must make sure that the weights or amounts of the foods being costed are equitable.

For example, a recipe requires 2 cups plain flour. 2 cups plain flour = 250 g (using weight conversion charts). Flour was purchased in a 2 kg bag costing $2.85. Use common weight amounts, that is, convert your purchase quantity to your required quantity. In this example, two kilograms were converted to grams.

$$\text{Flour cost} = \frac{\$3.60}{2000\text{ g}} \times 250\text{ g} = \$0.45$$

Therefore the cost of flour for the recipe was 45 cents.

Activity 5.7: Costing

1. What is the process for costing whole items such as fresh fruit or vegetables?

2. Work out the cost of a meal, using a table such as the one shown on the following page. You may need to use costing books from your classroom or websites that have food costs on them.

Ingredient	Original item cost	Quantity purchased	Recipe amount	Final cost
Diced pork			750 g	
Red apples			3	
New potatoes			6	
Carrot			1	
Zucchini			½	
Butter			15 g	
Brown sugar#			¼ cup	
Golden syrup#			1 T	
Apple juice			¼ cup	
Lemon rind*			½ tsp*	
Lemon juice*			2 T*	
Butter			60 g	
Caster sugar#			½ c	
Eggs			2	
Plain flour#			¼ c	
Cocoa powder#			1 T	
Milk fresh #			$^{2}/_{3}$ c	
Orange rind*			2 tsp*	
Orange juice*			2 T*	
Icing sugar#			1 tsp	
TOTAL MEAL COST				

* Sometimes easier to use a fresh ½ lemon or orange and cost accordingly.
# Convert cup and spoon measures to grams or millilitres.

NOTE: Use the weights and measures chart in Chapter 5 to convert cups and spoon measures to purchase weight. Write the weight next to the recipe amount.

WEIGHTS AND MEASURES OF COMMON FOODSTUFFS

The tables presented on the following pages will provide you with a quick reference for converting cups and spoons to grams. If the ingredient you are looking for is not listed, select a similarly weighted ingredient and use those amounts.

Note: These are approximate equivalents.

TABLE 5.1: Weights and measures of common foods

INGREDIENT	15 g	30 g	60 g	90 g	125 g	155 g	185 g	220 g	250 g
Almonds, ground	2 T	¼ C	½ C	¾ C	1¼ C	1⅓ C	1⅔ C	2 C	2¼ C
Almonds, slivered	6 t	¼ C	½ C	¾ C	1¼ C	1⅓ C	1⅔ C	2 C	2¼ C
Almonds, whole	2 T	¼ C	⅓ C	½ C	¾ C	1 C	1¼ C	1⅓ C	1½ C
Apples, dried, whole	3 T	½ C	1 C	1⅓ C	2 C	2⅓ C	2¾ C	3⅓ C	3¾ C
Apricots, dried, chopped	2 T	¼ C	½ C	¾ C	1 C	1¼ C	1½ C	1¾ C	2 C
Apricots, whole	2 T	3 T	½ C	⅔ C	1 C	1¼ C	1⅓ C	1½ C	1¾ C
Arrowroot	1 T	2 T	⅓ C	½ C	⅔ C	¾ C	1 C	1¼ C	1⅓ C
Baking powder	1 T	2 T	⅓ C	½ C	⅔ C	¾ C	1 C	1 C	1¼ C
Barley	1 T	2 T	¼ C	½ C	⅔ C	¾ C	1 C	1 C	1¼ C
Bicarbonate soda	1 T	2 T	¼ C	½ C	⅔ C	¾ C	1 C	1 C	1¼ C
Breadcrumbs, dry	2 T	¼ C	½ C	¾ C	1 C	1¼ C	1½ C	1¾ C	2 C
Breadcrumbs, soft	¼ C	½ C	1 C	1½ C	2 C	2½ C	3 C	3⅔ C	4¼ C
Biscuit crumbs	2 T	¼ C	½ C	¾ C	1¼ C	1⅓ C	1⅔ C	2 C	2¼ C
Butter	3 t	6 t	¼ C	⅓ C	½ C	⅔ C	¾ C	1 C	1 C
Cheese, grated, cheddar	6 t	¼ C	½ C	¾ C	1 C	1¼ C	1½ C	1¾ C	2 C
Cheese, parmesan, romano	6 t	¼ C	½ C	¾ C	1 C	1⅓ C	1⅔ C	2C	2¼ C
Cherries, glacé chopped	1T	2T	⅓ C	½ C	¾ C	1C	1C	1⅓ C	1½ C
Cherries, whole	1T	2T	⅓ C	½ C	⅔ C	¾ C	1C	1¼ C	1⅓ C
Cocoa	2 T	¼ C	½ C	¾ C	1¼ C	1⅓ C	1⅔ C	2 C	2¼ C
Coconut, desiccated	2 T	⅓ C	⅔ C	1 C	1⅓ C	1⅔C	2 C	2⅓ C	2⅔ C
Coconut, shredded	⅓ C	⅔ C	1¼ C	1¾ C	2½ C	3 C	3⅔ C	4⅓ C	5 C
Cornflour/Cornstarch	6 t	3 T	½ C	⅔ C	1 C	1¼ C	1½ C	1⅔ C	2 C
Coffee, ground	2 T	⅓ C	⅔ C	1 C	1⅓ C	1⅔ C	2 C	2⅓ C	2⅔ C
Coffee, instant	3 T	½ C	1 C	1⅓ C	1¾ C	2¼ C	2⅔ C	3 C	3½ C
Cornflakes	½ C	1 C	2 C	3 C	4¼ C	5¼ C	6¼ C	7⅓ C	8⅓ C
Cream of tartar	1 T	2 T	⅓ C	½ C	⅔ C	¾ C	1 C	1 C	1¼ C
Currants	1 T	2 T	⅓ C	⅔ C	¾ C	1 C	1¼ C	1½ C	1⅔ C

INGREDIENT	15 g	30 g	60 g	90 g	125 g	155 g	185 g	220 g	250 g
Custard powder	6 t	3 T	½ C	⅔ C	1 C	1¼ C	1½ C	1⅔ C	2 C
Dates, chopped	1 T	2 T	⅓ C	⅔ C	¾ C	1 C	1¼ C	1½ C	1⅔ C
Dates, whole, stoned	1 T	2 T	⅓ C	½ C	¾ C	1 C	1¼ C	1⅓	1½ C
Figs, dried, chopped	1 T	2 T	⅓ C	½ C	¾ C	1 C	1 C	1⅓ C	1½ C
Flour, plain, self raising	6 t	¼ C	½ C	¾ C	1 C	1¼ C	1½ C	1¾ C	2 C
Flour, wholemeal	6 t	3 T	½ C	⅔ C	1 C	1¼ C	1⅓ C	1⅔C	1¾ C
Fruit, dried, mixed	1 T	2 T	⅓ C	½ C	¾ C	1 C	1¼ C	1⅓ C	1½ C
Gelatine	5 t	2 T	⅓ C	½ C	¾ C	1 C	1 C	1¼ C	1½ C
Ginger, crystal, pieces	1 T	2 T	⅓ C	½ C	¾ C	1 C	1¼ C	1⅓ C	1½ C
Ginger, ground	6 t	⅓ C	½ C	¾ C	1¼ C	1½ C	1¾ C	2 C	2¼ C
Glucose, liquid	2 t	1 T	2 T	¼ C	⅓ C	½ C	½ C	⅔ C	⅔ C
Golden syrup	2 t	1 T	2 T	¼ C	⅓ C	½ C	½ C	⅔ C	⅔ C
Haricot beans	1 T	2 T	⅓ C	½ C	⅔C	¾ C	1 C	1 C	1¼ C
Honey	2 t	1 T	2 T	¼ C	⅓ C	½ C	½ C	⅔ C	⅔ C
Jam	2 t	1 T	2 T	¼ C	⅓ C	½ C	½ C	⅔ C	¾ C
Lentils	1 T	2 T	⅓ C	½ C	⅔C	¾ C	1C	1 C	1¼ C
Milk powder, fullcream	2 T	¼ C	½ C	¾ C	1¼ C	1⅓ C	1⅔ C	2 C	2¼ C
Milk powder, non-fat	2 T	⅓C	¾ C	1¼ C	1½ C	2 C	2⅓ C	2¾ C	3¼ C
Nutmeg, ground	6 t	3 T	½ C	⅔ C	¾ C	1 C	1¼ C	1½ C	1⅔ C
Nuts, chopped	6 t	¼ C	½ C	¾ C	1 C	1¼ C	1 ½ C	1 ¾ C	2 C
Oatmeal	1 T	2 T	½ C	⅔ C	¾ C	1 C	1¼ C	1½ C	1⅔ C
Olives, whole	1 T	2 T	⅓ C	⅔ C	¾ C	1 C	1¼ C	1 ½ C	1⅔ C
Olives, sliced	1 T	2 T	⅓ C	⅔ C	¾ C	1 C	1¼ C	1½ C	1⅔ C
Pasta, short	1 T	2 T	⅓ C	⅔ C	¾ C	1 C	1¼ C	1½ C	1⅔ C
Peaches, dried whole	1 T	2T	⅓ C	⅔ C	¾ C	1 C	1¼ C	1½ C	1⅔ C
Peaches, chopped	6 t	¼ C	½ C	¾ C	1 C	1¼ C	1½ C	1¾ C	2 C
Peanuts, raw, whole	1 T	2 T	⅓ C	½ C	¾ C	1 C	1¼ C	1⅓ C	1½ C
Peanuts, roasted	1 T	2 T	⅓ C	⅔C	¾ C	1 C	1¼ C	1½ C	1⅔ C
Peanut butter	3 t	6 t	3 T	⅓ C	½ C	½ C	⅔C	¾ C	1 C
Peas, split	1 T	2 T	⅓ C	½ C	⅔C	¾ C	1 C	1 C	1¼ C
Peel, candied, mixed	1 T	2 T	⅓ C	½ C	¾ C	1 C	1 C	1¼ C	1½ C
Potato, powder	1 T	2 T	¼ C	⅓ C	½ C	⅔ C	¾ C	1 C	1¼ C
Potato, flakes	¼ C	½ C	1 C	1⅓ C	2 C	2⅓ C	2¾ C	3⅓ C	3¾ C
Prunes, chopped	1 T	2 T	⅓ C	½ C	⅔C	¾ C	1 C	1¼ C	1⅓ C
Prunes, whole, stoned	1 T	2 T	⅓ C	½ C	⅔C	¾ C	1 C	1 C	1¼ C
Raisins	2 T	¼ C	⅓ C	½ C	¾ C	1 C	1 C	1⅓ C	1½ C

INGREDIENT	15 g	30 g	60 g	90 g	125 g	155 g	185 g	220 g	250 g
Rice, short grain	1 T	2 T	¼ C	½ C	2/3 C	¾ C	1 C	1 C	1¼ C
Rice, long grain	1 T	2 T	1/3 C	½ C	¾ C	1 C	1¼ C	1 1/3 C	1½ C
Rice bubbles	2/3 C	1¼ C	2½ C	3 2/3 C	5 C	6¼ C	7½ C	8¾ C	10 C
Rolled oats	2 T	1/3 C	2/3 C	1 C	1 1/3 C	1¾ C	2 C	2½ C	2¾ C
Sago	2 T	¼ C	1/3 C	½ C	¾ C	1 C	1 C	1¼ C	1½ C
Salt, common	3 t	6 t	¼ C	1/3 C	½ C	2/3 C	¾ C	1 C	1 C
Semolina	1 T	2T	1/3 C	½ C	¾ C	1 C	1 C	1 1/3 C	1½ C
Spices	6 t	3 T	¼ C	1/3 C	½ C	½ C	2/3 C	¾ C	1 C
Sugar, granulated	3 t	6 t	¼ C	1/3 C	½ C	2/3 C	¾ C	1 C	1 C
Sugar, caster	3 t	5 t	¼ C	1/3 C	½ C	2/3 C	¾ C	1C	1¼ C
Sugar, icing	1 T	2 T	1/3 C	½ C	¾ C	1 C	1C	1¼ C	1½ C
Sugar, brown	1 T	2 T	1/3 C	½ C	¾ C	1 C	1C	1 1/3 C	1½ C
Sultanas	1 T	2 T	1/3 C	½ C	¾ C	1 C	1C	1¼ C	1½ C
Tapioca	1 T	2 T	1/3 C	½ C	2/3 C	¾ C	1C	1¼ C	1 1/3 C
Treacle	2 t	1 T	2T	¼ C	1/3 C	½ C	½ C	2/3 C	2/3 C
Walnuts, chopped	2 T	¼ C	½ C	¾ C	1 C	1¼ C	1½ C	1¾ C	2 C
Walnuts, halved	2 T	1/3 C	2/3 C	1 C	1¼ C	1½ C	1¾ C	2¼ C	2½ C
Yeast, dried	6 t	3 T	½ C	2/3 C	1C	1¼ C	1 1/3 C	1 2/3 C	1¾ C
Yeast, compressed	3 t	6 t	3 T	1/3 C	1/3 C	½ C	2/3 C	¾ C	1 C

Egg sizes used are of average 60 g, unless specified in recipes.

Useful Websites

www2.woolworthsonline.com.au/: Woolworths online ordering system. Pricing and product selection

http://shop.coles.com.au/online/national/: Coles also operates an ordering system with prices available on their website

http://www.taste.com.au/how+to/articles/369/weights+measurement+charts: Information on Australian weights and measures

www.chefpedia.org/wiki/index.php?title=Culinary_Terminology: A comprehensive list of culinary terms

CHAPTER 6
Making it Work

Key Concepts

- Working to prepare food
- Working in a team
- Communication and collaboration
- Workflow planning
- Production planning and production proposals
- Workflow efficiency
  - personal efficiency
- Kitchen efficiency
- Food preparation efficiency
- Food preparation practices
  - kitchen safety
- First aid
- Hygiene practices
  - personal hygiene
  - food hygiene

WORKING TO PREPARE FOOD

Whether you are in the kitchen at home, the food science rooms at school or even in a workplace preparing food there are behaviours and expectations that should be met in varying degrees depending on the environment. Whilst it may be permissible at home to lick the spoon after preparing a dish, this sort of behaviour is not acceptable when preparing food with and/or for others. Other personal habits, traits and hygiene practices all come under the microscope in the food preparation environment. So what makes a good food handler? Have you watched someone working with food and admired their habits, practices and skills or have you been disappointed and put off by some other food handler's behaviours.

Activity 6.1: You and food handling

1. Describe the attributes you would like a person to have if they were to prepare a meal for you.
2. Think about yourself and list what qualities you possess that would make you a suitable food handler.
3. Identify an area that you could work to improve upon to make yourself a more proficient food handler. Explain how you will accomplish this improvement.

WORKING IN A TEAM

This is a fundamental aspect of working in a food environment and is not limited to the workplace. How you relate to others and your confidence with them as well as basic communication skills establish the foundations of what is to come. Rarely does a good end product (and hence chef satisfaction) come from a disharmonious working environment. Teamwork is something that needs to be strived for, maintained and continually evaluated to seek improvement. It involves intertwining the skills outlined below and needs to be applied by all participants.

When dealing with others it is important to remember that first impressions are lasting and that professional courtesy must be used at all times. A happy working environment where respect is shown for all has proven to achieve higher productivity than those workplaces with a less satisfactory working environment.

Some successful communication ideas:

- Always respect and support others.
- Be aware of the importance of body language.
- Be careful to avoid conflict.
- Be friendly, yet professional.
- Listen to and consider other people's opinions.
- Make sure others understand what you are trying to say.
- Recognise and respect cultural differences.
- Think before you speak.
- Try not to use slang or double meanings.
- Use active listening techniques.
- Keep instructions clear and concise.

COMMUNICATION AND COLLABORATION

Whenever more than two people work together, a set of roles for each person is developed, whether verbally or by natural group processes. Once the goal that you are trying to achieve is established, the group members will work out what each person has to do, often delegating tasks that each person has competence or expertise in so that the job will be achieved to the best of the group's ability.

WORKFLOW PLANNING

Another aspect of kitchen and food preparation management is that of workflow planning, otherwise known as time plans, work plans or management plans. A good workflow document outlines the steps in preparation (method) combined with the foods required (ingredients) and the items needed to prepare it (utensils and equipment). It may also incorporate a timing component so that the dish is prepared by a specific time (that is, for serving or presentation). A sample format is outlined on the next page.

In a commercial kitchen, or even when you are at home or at school, simultaneously preparing two dishes, a type of workflow planning is essential to ensure proceedings run as smoothly as possible. Commercially, a chef will need to know who is responsible for each dish required.

Basically a workflow plan is information set out in a logical sequence so that tasks are completed on time. A workflow plan outlines details such as what, when and how. It is in a little more detail than a recipe and is useful when working with limited time when you have to be really organised and be able to follow a strict time plan.

A workflow plan:

- saves time
- minimises food wastage
- increases productivity
- increases organisation
- raises awareness of safety issues.

A typical workflow plan is outlined on the following page.

Interpersonal skills are attributes that enable a person to successfully communicate and work with other people to provide a happy and productive outcome. Initiative is an example of an interpersonal skill.

Organisational skills are very important in food management and require a person to ensure their work practices run smoothly by making sure all of their needs are met. This involves being aware of what needs to be done when, with what and by whom, and ensuring all resources to complete the task are on hand. It also entails the management of the workspace to create an environment conducive to productivity.

Communication skills are skills that can be verbal, written and non-verbal and involve a person providing information to others.

Teamwork involves individuals working together successfully to achieve a common goal. Teamwork skills include a combination of interactive, interpersonal, problem solving and communication skills required by a group of people working on a common task in a complementary manner.

Recipe/Meal:		**Date:** (day and date)	**Time:** (class time)	**Source:** (required for cross referencing)
Time (in 5-10 minute intervals)	**Ingredients** (what and how much)	**Equipment** (utensils as well as small and large equipment)	**Techniques** (key processes and techniques and also who does what)	**Safety** (considerations)

Hint: when working out the times for completing a recipe for a specific serving time, it is easier to work backwards. For example, to prepare a cake, start off with its cool-down and final presentation time, then the baking time, then the preparation time and finally the initial organisation and set-up time. This will ensure that you keep to your serving time without the cake not being cooked or, alternatively, being ready too early. Allow yourself a few extra spare minutes before serving to allow for any unforseen problems that may cause you to run late (such as equipment failure).

For example – the recipe presented on the following page is set out in the format of a workflow plan and is to be ready by 10.00 am.

Activity 6.2: Workflow planning

1. Use the workflow plan outlined above to prepare the Lemon Pumpkin Scones as outlined, changing times to suit your class time. What timing issues did you have if any?

   Were there any errors in the workflow plan that you found?

   Did you find the workflow plan useful or are there modifications that you could make?

2. Use the workflow format of the workflow plan outlined previously to plan and prepare a recipe selected from the recipe chapter included in this book to serve at the end of one of your cooking lessons. Fill in the details using a table such as the one above. Starting time will be the start of your lesson.

Sample recipe

Recipe/Meal:	Date:	Time:	Source:
Lemon Pumpkin Scones	Thursday 2 March	Period 1	Women's Weekly Menu Planner No.12

Time	Ingredients	Equipment	Techniques	Safety
9.00			Personal preparation. Preheat oven 180°C.	Oven safety
9.05			Collect ingredients.	
9.10	60 g butter ¼ c caster sugar	Electric mixer Medium bowl Measuring scales Measuring cups	Cream butter and sugar until light and fluffy.	Electrical safety
9.14	1 egg, beaten lightly	Electric mixer Small bowl Fork	Beat in egg to combine.	Electrical safety
9.16	1 c mashed pumpkin, drained 2½ c SR flour 1 Tbsp milk, approx	Large bowl Table knife Measuring cups Measuring spoons	Transfer to bowl. Stir in pumpkin, half flour and mix with knife. Add rest of flour and milk if needed for a soft dough.	
9.20	½ c chopped mixed peel	Greased oven tray Rolling pin Table knife Chef's knife Measuring cups	Roll out dough. Sprinkle with peel and roll up. Place on tray to form a ring. Cut outside edge at 4 cm intervals into the centre.	Knife safety
9.30		Oven	Bake in oven.	Oven safety
9.50		Wire rack	Remove from oven.	Oven safety
9.52	1½ c icing sugar 1 tsp soft butter 2 Tbsp lemon juice	Sieve Heatproof bowl Saucepan hot water Wooden spoon Spatula	Sift icing sugar into bowl. Add butter and lemon juice. Work to stiff paste over hot water.	Stovetop safety
9.57	Extra grated lemon rind	Grater	Ice scones. Decorate with rind.	
10.00		Serving plate	Serve.	

PRODUCTION PLANNING AND PRODUCTION PROPOSALS

A **Production Proposal** in a cooking situation refers to an outline (even a summary) of a cooking event that details the time, place, event, clients, costing and logistics. This is often a method used when 'tendering' or vying for a task or job that someone else wants done. For example, if your class was asked to cater for a teachers' breakfast within a set budget, you could implement a production proposal to give to the planners of the breakfast prior to its implementation. That way, you can plan the details such as kitchen staff, waitpersons, food deliveries and costs covered. It is written in dot points or short headings and is not very long or detailed.

Aspects of a Production Proposal could include:

- day and time
- venue or place
- number of guests or clients
- menu outline
- approximate costs involved
- staff required
- logistics eg. – prior preparation; delivery timings; serving constraints if any
- special dietary needs for consideration if any.

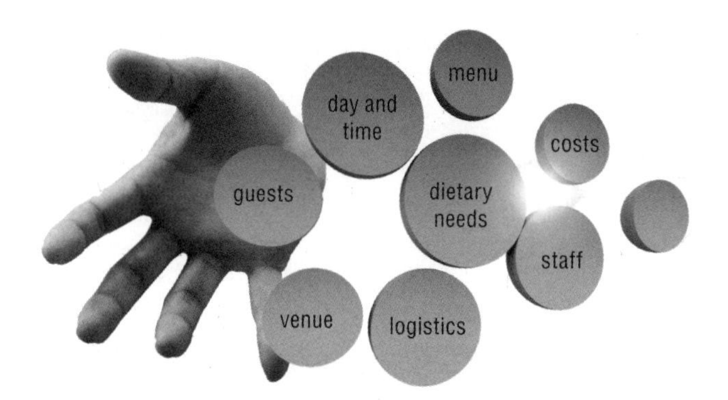

Activity 6.3: Production proposal

Devise a production proposal based around the following information.

The cooking groups within your class have been invited to tender for the catering of the Phys Ed Department's athletics day lunch. You will need to supply food to the 35 athletes competing plus the five staff at Perry Lakes Stadium. It is not required to include drinks. Think about the above dot points when you create your production proposal. You will have the classroom to use for preparation.

Production planning

Production planning tends to be a little more complicated than the workflow plan and production proposal. A **production plan** assists in the management of creating more than one recipe in order to meet a specific time frame. In a commercial kitchen, or even when you are at home or at school, simultaneously preparing two dishes, some type of production planning is required to ensure proceedings run smoothly and the food is served when required.

A production plan is also beneficial when small groups of two to three people are working together to produce a multiple course meal. It allocates tasks with specific ingredients to individuals and should hopefully ensure that all dishes come together at the appropriate serving time.

A production plan:

- saves time
- minimises food wastage
- increases productivity
- increases organisation
- raises awareness of safety issues
- enables a group to work together successfully.

Activity 6.4: Production planning

1. Using the workflow plan as a basis, develop a production plan format that would suit you and your cooking group to work with. You will need to incorporate a section for who does what task. Outline this format.

2. With your developed format, complete a production plan for the following two recipes to be served simultaneously to your teacher. Don't forget to include steps to cook the carrots, potato and zucchini. Complete the preparation task and evaluate your plan. How would you have to modify it to improve your performance?

Pork and Apple Kebabs

750 g diced pork
3 red apples, chopped
6 new potatoes
1 carrot, julienne
½ zucchini, julienne

Marinade:

15 g butter
1 Tbsp golden syrup
½ tsp grated lemon rind
¼ c brown sugar
¼ c apple juice
2 Tbsp lemon juice

1. Combine pork and marinade in large bowl, cover, refrigerate 2 hours.
2. Drain pork, reserve marinade.
3. Thread pork and apple alternately onto bamboo skewers.

4. Grill for about 6 minutes on each side. Brush occasionally with reserved marinade.
5. Serve glazed with any remaining marinade accompanied with steamed carrot and zucchini and baked potatoes.

Chocolate Citrus Puddings

60 g butter
½ c castor sugar
2 eggs, separated
¼ c plain flour
1 Tbsp cocoa
⅔ c milk
2 tsp grated orange rind
2 Tbsp orange juice
icing sugar

1. Cream butter and sugar in small bowl with beater until combined.
2. Beat in egg yolks, sifted flour and cocoa.
3. Gradually add milk, rind and juice, and mix until smooth.
4. Beat egg whites in small bowl with beater until firm peaks form.
5. Fold egg whites through chocolate mixture.
6. Pour into 4 greased ovenproof dishes (½ cup capacity), place dishes onto a baking tray, bake in moderate oven for about 20 minutes or until cooked.
7. Serve dusted with icing sugar.

WORKFLOW EFFICIENCY

Workflow efficiency is defined as making a task easy and quick to complete by using simplified movements, techniques and processes.

A kitchen can be a stressful environment – especially in those kitchens catering to the dining trade. Economics make it feasible to often have more dining space compared to preparation and storage space. Eating habits also are designated to certain periods of the day, making peak times in the kitchen a very busy situation in an often smaller than desired area in hot working conditions! Therefore a kitchen needs to be efficient – in terms of workspace and the people working in it. Domestically, kitchens are designed for one to two persons but even then, it can become a crowded and dangerous space if not planned correctly!

PERSONAL EFFICIENCY

Personal efficiency refers to how a person goes about their work and emphasis is put on using quality movement and time to complete the task.

Try to incorporate personal habits into food preparation efficiency. Hints on how to do this include:

- Make sure you follow a set plan.
- Ensure all stock and equipment that you need is on hand prior to the event.

- Maintain your equipment well; for example, knives must be sharp.
- Try to do some tasks well beforehand, that is, mis en place preparation or preparation steps. Some recipes outline what parts can be done prior to the final preparation.
- Make tasks easier or simpler where you can.
- Attempt to dovetail tasks or combine tasks together.
- Economise all movements; this comes with organisation.
- Ask for help from others when required.
- Do not let yourself become too fatigued or tired.
- Dress comfortably, especially footwear.

KITCHEN EFFICIENCY

For a kitchen to run successfully it needs to be planned carefully. This helps prevent accidents and injuries from people having to move around a lot, or having the equipment and materials in awkward places. In a commercial situation, a kitchen needs to follow the pathway of the food preparation process.

Think about your kitchen at home. Is the kitchen located close to where groceries are brought into the house, for example the front door? Do you have to go up or down stairs with heavy bags of groceries? Is your main dining area close to the kitchen? What if you have a meal outside, is this area close?

Activity 6.5: Kitchen workflow

1. Describe the efficiencies and inefficiencies of your kitchen workflow at home. Draw a diagram of its layout. What can be done to improve it? Make recommendations if you were to redesign your kitchen to make it more efficient.

2. To test the efficiency of your kitchen, prepare a chocolate cake. Have someone note your movements during the process and together come up with ideas that could improve the workflow and kitchen efficiency.

FOOD PREPARATION EFFICIENCY

It is often interesting to watch someone preparing a meal. If they are not concentrating, the entire process may take a lot longer as they forget what they are doing, change their mind about things or simply don't know where to start! If in a hurry, it is mindful to use some of the personal efficiency hints mentioned previously as well as the ones listed below.

- Work within a small area to minimise movement.
- Keep most frequently used items close at hand.
- Work left to right, (or right to left if you are left-handed).
- Centre your chopping board with foods on one side, a tray for processed foods on the other and a scraps bowl nearby.
- When preparing bulk foods, deal with one process at a time. For example, deal with all of the peeling of carrots, then the trimming, then the final cutting.
- Always keep your workbench clear when possible. Put utensils away and clear the scraps regularly.

Activity 6.6: Observation

Watch a colleague preparing foods, taking notes on their personal inefficiencies as they go. At the end of the food preparation, discuss their work habits and provide ideas for where they can improve their personal efficiency.

FOOD PREPARATION PRACTICES

When preparing food in any kitchen, safety needs to be addressed at all times, including both personal safety and food safety. At the end of the day you want everybody involved in the food preparation process to be healthy and injury free. This involves keeping the food safe to eat and the kitchen safe to work in.

Kitchen safety

Any environment has its potential for injury and the kitchen is a prime example of this – whether it is a domestic home kitchen or a foods room at school. There are certain elements that cause safety concerns and in kitchens they are primarily heat, water, sharp objects and people plus a variety of other smaller concerns. An awareness of these safety issues can often prevent injuries from happening.

1. Heat

Burns and scalds can be caused from hot equipment, food or liquids. Care needs to be taken when working with ovens and cook tops or equipment that prepares hot foods. Technology has improved some domestic equipment with heatproof materials; however, commercial kitchens still utilise metals that retain their heat.

2. Water

This has elements of scalding if too hot and spills on floors can cause potential injury from falls. As well, its combination with electricity can produce lethal effects. Steam is also a hazard.

3. Machinery and tools

The kitchen has a lot of equipment and tools that are sharp and/or have moving parts. These elements can make a dangerous environment for the unwary. Knives and items with blades are especially a hazard.

4. Work environment

This covers aspects such as ventilation, floor surfaces and floor layout. Injuries can occur through heat stress and slipping or tripping on objects on the floor. If the premises are also not ergonomically set out, collisions between workers can occur.

5. Manual handling

This involves using your body to carry, bend, lift and stretch. If incorrect posture is used for manual handling it can cause muscle strain and/or injury.

6. Electricity

Electrical appliances bring their own hazards into the kitchen because electric shocks can be fatal. Combined with the presence of water, electricity is a serious safety concern and should not be ignored. Equipment with frayed cords or faulty switches should be repaired before using.

7. Gas

While it is a great source of energy, if mismanaged gas can cause leaks, with potential explosions and burns from the flames.

8. Fire

Some kitchens operate with an open or enclosed fire as a source of heat for cooking,creating a multitude of safety concerns. Fire can also be caused by overheating oil, or faulty electrical equipment.

Activity 6.7: Safe food preparation practices

1. For two of the issues outlined above, create a safety management plan outlining potential hazards and what can be done to minimise their risk of injury. Discuss your responses with your class members.

2. Try to think up a few more ideas on personal efficiency and list these.

FIRST AID

Having knowledge of first aid is a necessary part of any practical course. With quick action, injuries can be managed effectively to minimise complications, such as further bleeding or choking. The basic first aid procedure follows the plan known as **DRABC** outlined on the following page.

For further information regarding EAR (expired air resuscitation) and CPR (cardiopulmonary resuscitation) look up your nearest First Aid agency such as St John Ambulance or Red Cross organisation to find out about their first aid courses. Completion of such a qualification is helpful to you personally as well as an advantage when seeking a job. Your school nurse may also be of assistance regarding first aid information.

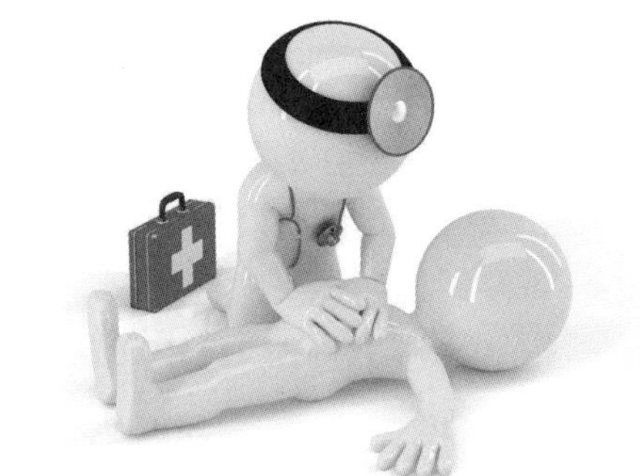

For injuries that are less life threatening, simple management procedures can be taken. Make sure you are careful about the exchange of bodily fluids and be prepared to wear disposable gloves when dealing with other persons for their health and safety as well as yours. Generally, if possible, it is safer for a person to deal with their own blood themselves.

HYGIENE PRACTICES

A lot of awareness has developed over the years regarding food hygiene both in the home and in food production areas. With the emergence of technological processes such as refrigeration and pasteurisation as well as the occurrence of food being partially or fully prepared outside the home, people's expectations are for safe quality foods.

Food can become easily contaminated therefore anyone involved in the preparation of food must realise their responsibility in keeping the food safe for them and others to eat. Food contaminants come from a variety of sources and contamination can happen during the initial growing stages, transportation, storage, point of sale, preparation and/or serving of the ingredients.

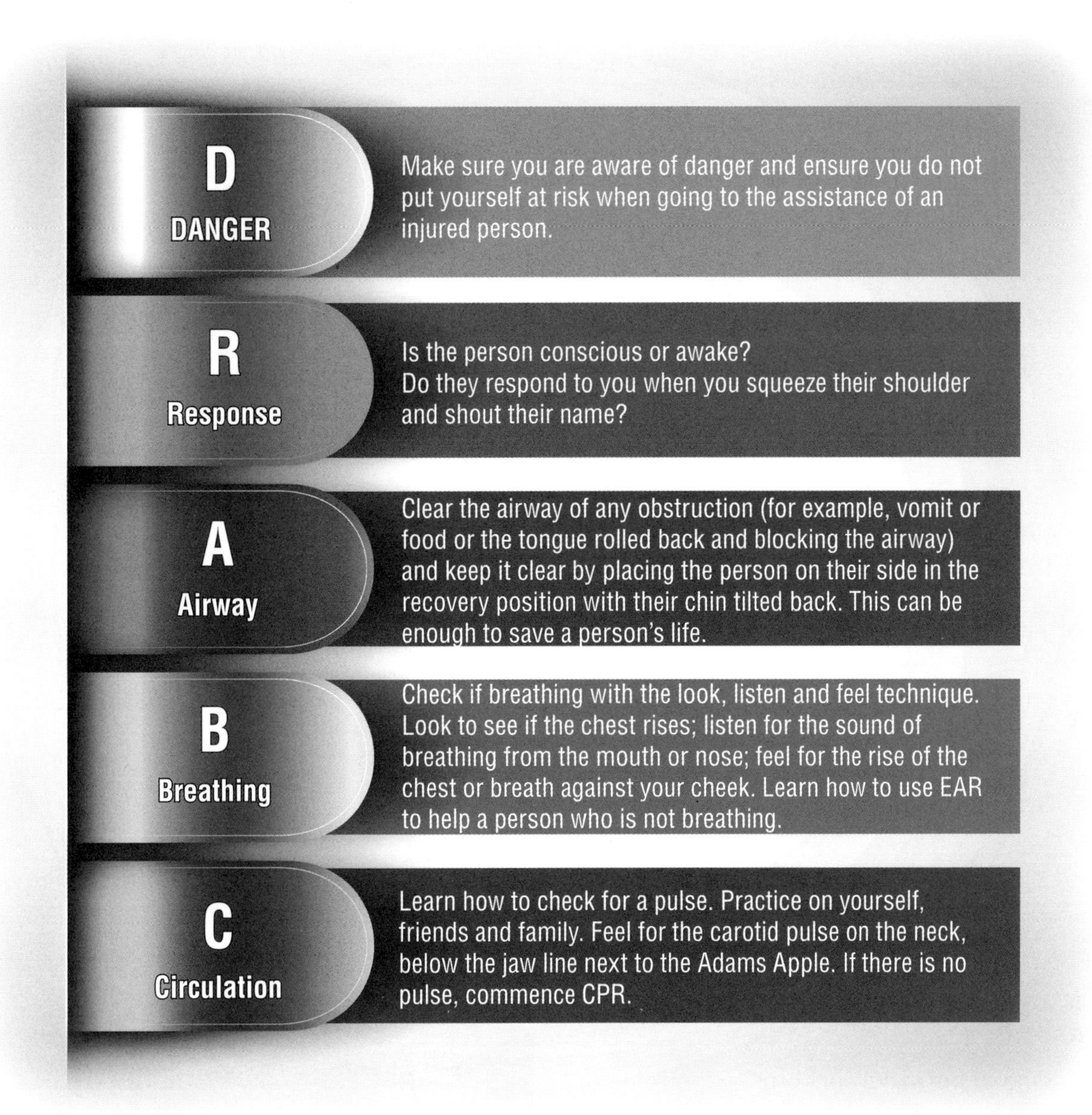

Domestically, people are becoming more and more aware of how to prevent food spoilage and contamination in the home and, with the advent of better storage and cleaning products/solutions, this is becoming an easier task than that faced by our ancestors.

The presence of harmful bacteria in food is the main cause of food poisoning. While some bacteria can be processed readily by our bodies, there are others that cause harm, even death, to humans.

Contamination is the pollution or infection of a substance with foreign matter.

Bacterium is a kind of microscopic organism that exists widely in nature, some of which cause disease and decay.

Did you know ...

Gelatine is one of the foods identified as being the most likely to cause food poisoning.

Activity 6.8: Hygiene

1. Find out about the Food Safety Temperature Zone and create your own thermometer guide.
2. Using the diagram you have created, explain what could happen if you reheat lukewarm leftovers that have not been refrigerated.
3. Find out the name of four types of harmful bacteria and the foods in which they are most likely to occur.

Whilst cooked food has had a lot of bacteria destroyed from the application of heat, raw food still has these bacteria present which creates issues if either comes into contact with each other.

Cross-contamination of food involves the transfer of harmful bacteria from uncooked or raw food to food which has already been cooked or prepared.

Did you know ...

Honey is the only natural food that does not spoil. It was found in the tombs of Egyptian Pharaohs and was edible!

Personal hygiene

One of the initial concerns of food safety is ensuring that individuals handling food follow basic safe food handling practices. This involves ensuring your personal hygiene, cleanliness and behaviour with food is performed to industry standards when preparing food for others. Bacteria are present on your body, hair, clothing and jewellery and can easily be spread onto the food that you are preparing if you are not careful.

Because a food worker's hands are generally what come into contact most frequently with ingredients, it is important to understand when it is necessary to wash your hands.

Activity 6.9: Personal hygiene

1. Complete a list of practices you would like to see individuals perform when preparing food for others. Two examples are given to start your list.
   - Tie your hair back or cover it with a hairnet.
   - Do not work if you are sick.
2. With your class, list a range of situations which would necessitate the washing of hands during the process of preparing food.
3. Explain why a butcher would have to be particularly careful when working and serving customers?

Food hygiene

As well as being aware of the importance of personal hygiene, the actual foodstuffs themselves need to be taken care of to ensure they do not become contaminated through other means. There are many aspects that contribute to food contamination, such as storage and exposure to moisture.

Food spoilage refers to the natural deterioration of a food product through time, or damage which affects its flavour, texture and appearance.

Activity 6.10: Food hygiene

1. What other aspects can you think of that could contribute to food becoming contaminated?
2. You have purchased a box of granny smith apples from the orchard cheaply. In order to make the most of your purchase how could you manage the apples to prevent food spoilage and avoid wastage?

EXTENSION

Extension Activities

1. Create an experiment to see how hygienic you are! You may need help from the science department at your school. Develop a bacterial culture using an agar solution taking swabs from your hands, your hair and your mouth (maybe sneezing). Allow it to develop over a few days and see what happens. There will be precautions that need to be taken as you are developing a biohazard material. What does this experiment tell you regarding food and personal hygiene? If you were to work in the food industry what must you ensure that you do to keep the food you prepare safe to eat?
2. Leave a variety of foods uncovered out on the bench for a few days. Have fresh foods as well as processed foods. Describe the changes in the foodstuffs over a week and make recommendations for storage. What were the differences of those foods that contained preservatives? Has this experiment changed your views on storage of food and, if so, how? As another element, store some foods in the fridge uncovered to compare what happens.
3. Compare the production of a freshly home-made meal compared to a takeaway food. Look at factors such as cost, production time, ease of preparation, ingredients used, etc. What are the unforeseen contamination risks to both end products (that is, food contamination opportunities before your use or purchase of the food item). Make recommendations on your findings.

Useful Websites

www.beefandlamb.com.au/How to/Buying beef and lamb: Contains information about keeping meat safe to eat.

www.foodsafety.asn.au: Australia's leading body on helping families to prevent illness through food contamination.

www.foodstandards.gov.au: Information on food safety standards.

www.public.health.wa.gov.au/2/1061/2/food.pm: Provides procedures for reporting food poisoning and other food

www.foodsafety.com.au/news/: Up to date articles on food safety for Australians

CHAPTER 7
Getting Back to Basics

Key Concepts

- Tools and equipment
- Measuring in the kitchen
- Knives
- Cutting food
- Washing and cleaning fruits and vegetables
- Basic food preparation techniques
- Basic mis en place preparations

TOOLS AND EQUIPMENT

A good food production person knows how important it is to understand and look after their equipment. Using the right equipment for the job is vital if food is to be prepared and presented to a high standard. Hundreds (and sometimes) thousands of dollars can be spent on outfitting a domestic kitchen, let alone a commercial one – but is all of it necessary? Let's look at what different types of equipment are found in the kitchen, and the items that may require special care or handling.

Utensils

Utensils are items that are hand held and operated so are generally small in size. They vary in complexity and usage so a selection is listed below for you to investigate.

Activity 7.1: Utensils

1. The first is done for you, so follow the example and complete a description of the other utensils below.

   - **Balloon whisk:** a wire or plastic device made up of looping wires that is used to mix or beat mixtures to incorporate air into the final product.

   - Bread knife
   - Chef's knife
   - Chopsticks
   - Corer
   - Ladle
   - Masher
   - Measuring jug
   - Measuring spoons
   - Paring knife
   - Parisienne cutter
   - Perforated spoon
   - Sieve
   - Spatula
   - Vegetable peeler
   - Wooden spoon
   - Zester

2. What care needs to be taken when storing utensils? Think of hygiene and safety in your response and give examples where appropriate.

3. What special care needs to be taken for caring and storing knives? Explain your answer.

Small equipment

Generally, these are items that are larger than utensils.

Activity 7.2: Small equipment

The first is done for you, so follow the example and complete a description of the other items below.

- **Baking dishes**: metal or ceramic dishes that can contain food or mixtures to cook in the oven.

- Bamboo steamer
- Bowls
- Chinois/Conical strainer
- Chopping board
- Colander
- Funnel
- Grater
- Juicer
- Mandolin
- Measuring scales
- Mouli
- Pestle and mortar
- Rolling pin
- Saucepans
- Sieve
- Wok

Mechanical equipment

These are items that require electricity to operate and can be large or small in size.

Activity 7.3: Mechanical equipment

1. The first is done for you, so follow the example and complete a description of the other utensils below.

   - **Blender:** a lidded glass or plastic container with a set of short blades fixed into the base and powered by electricity to chop and mix at high speeds.

   - Coffee percolator
   - Electric knife
   - Food processor
   - Frypan
   - Hand held beater
   - Microwave oven
   - Rice cooker
   - Sandwich press
   - Slicer
   - Toaster

2. Think about your kitchen at home. List the mechanical equipment in your kitchen used for food preparation. For each of these items, rate their necessity to your lifestyle. If not entirely necessary, make comments as to what other equipment could be utilised in its place.

3. Look at an electrical supplier's catalogue. Create a list of the kitchen equipment shown as vital, useful and a luxury (not necessary). Justify your inclusions in each category.

Looking after equipment

As with any tools, if looked after properly kitchen equipment can last you a long time. Because they are in contact with variables such as acid, water and/or heat, if not correctly maintained they will deteriorate quickly and succumb to rust and corrosion as well as general wear and tear. Always follow the manufacturer's instructions.

Cleaning

All utensils and items that are non-electrical need to be washed in hot water and detergent. Use scourers or brushes to assist the removal of food particles or dirt. Ensure all items are rinsed in hot water and dried completely before storing.

Cleaning process

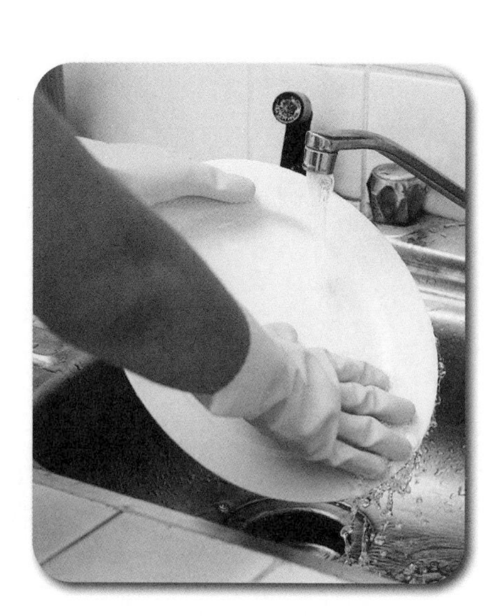

1. Scrape all excess food off utensils and equipment.
2. Fill a sink with hot water and add sufficient detergent. Too much requires extra rinsing and is damaging to our environment and water supplies. Too little will not clean the dishes properly.
3. Fill a second sink if available with hot water for rinsing. If no second sink is available leave dishes on the draining board and pour over hot water at the end of washing to remove detergent.
4. Wash glassware and rinse thoroughly. Dry stemware immediately with a lint free cloth.
5. Wash eating implements (cutlery) and rinse.
6. Follow with eating plates or crockery, rinsing well removing streaks.
7. Wash cooking utensils and small equipment.
8. Finally finish with pots and saucepans. This leaves the dirtiest dishes to the end allowing the cleanest water for the drinking and eating implements which are in more direct contact with the mouth when eating.

MATERIALS IN THE KITCHEN

When selecting equipment, there may be options available to you regarding what material is used in the equipment/utensil. Below are some more common materials used in the kitchen and their care required which may assist you in making decisions about purchasing equipment.

Wood

Wood is often used in chopping boards and spoons; however, wood harbours bacteria due to its porous nature. Because of this, wooden utensils are not recommended for food preparation. If wood is used, it should not be left soaking in water, it needs to be scrubbed with a brush when washed in hot water, and needs to dry thoroughly. Wood also absorbs colours and flavours so use separate wooden spoons for sweet and savoury dishes.

Plastic

Plastic comes in many different forms and is a common material in kitchens today. Most plasticware is dishwasher proof and easy to clean due to a non-porous surface. However, over time plastic scratches and perforates causing discolouration and deterioration. When this occurs it is wise to discard the affected item and replace it.

Some plastics deteriorate at high temperatures so check with the manufacturer before using plastic not for its original intended use (for example, some plastic containers melt in the microwave oven).

Metal

This is a favoured material for commercial cookery due to its neat, hygienic and easy to care for surface. It can be subjected to wear and tear and yet lasts a lot longer than some of its counterparts. Its conductivity makes it useful for cookware (saucepans) and progressive technology is producing metals that do not conduct heat (useful for saucepan handles).

Copper and aluminium have to be used with care due to their oxidising nature when in contact with acidic foods. Other more suitable metals such as stainless steel are more appropriate for heating foods.

Ceramic (including china and porcelain)

This is one of the earliest forms of kitchenware that was manufactured by man. It provides a smooth surface and can be formed into any shape required. It can be heavy, is fragile and care needs to be taken when handling it. Newer forms can be used to heat, serve and freeze, showing ceramic's versatility.

Activity 7.4: Cooking equipment materials

1. What other materials can you identify that are used in kitchen equipment in your home? What are their uses and limitations?
2. Explain the use of Teflon® on cooking surfaces and any benefits or limitations it has had.
3. What materials were most often used by your parents and their parents in the kitchen? Have there been any changes that they can identify?

MEASURING IN THE KITCHEN

Even though this appears to be a quite simple process, different interpretations of measurement abbreviations and different equipment sizing creates problems and issues when following recipes.

In Australia, recipes should be written incorporating measurements 'Approved to Australian Standards' or more commonly referred to as Australian; for example, one Australian metric cup.

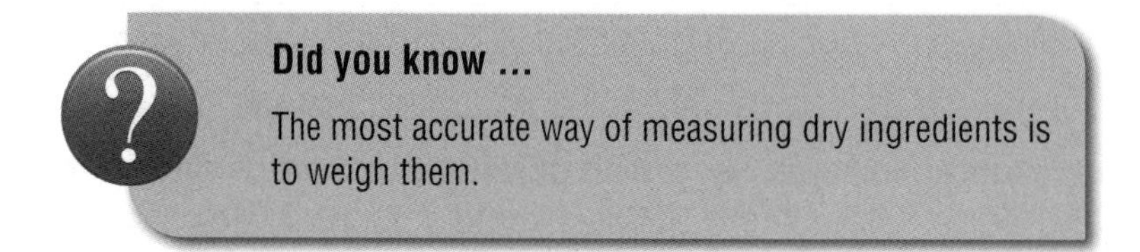

Did you know ...

The most accurate way of measuring dry ingredients is to weigh them.

Measuring abbreviations

These are the most common abbreviations that you will come across.

c, C	=	cup
t, tsp, ts	=	teaspoon
T, Tbsp, Tb	=	tablespoon
ml	=	millilitre
L	=	litrew
g, gm, gms	=	grams

kg	=	kilogram
mm	=	millimetre
cm	=	centimetre
oz	=	ounces (imperial)
lb	=	pound (imperial)

Measuring equipment

Note: These are Australian measures.

(Liquids)

1 metric cup	=	250 millilitres (ml)
1 litre	=	1000 ml, or 4 metric cups

(Liquid and dry)

1 tablespoon	=	20 ml
1 teaspoon	=	5 ml
½ teaspoon	=	2.5 ml
¼ teaspoon	=	1.25 ml

Measuring techniques

Follow these steps to ensure ingredient quantities are accurate.

Dry ingredients:

1. Place a measuring cup on a flat bench with paper underneath it.
2. Spoon in the ingredients until the measure is slightly overfull.
3. Using the back of a knife, flatten off and return the excess back to the original container, using the paper.

Liquid ingredients:

1. Place a clear jug or measure onto a level table.
2. Pour into the measure and check at eye level.

Note: Ingredients such as brown sugar, compressed yeast and softened butter should be lightly packed into the cup or measure, unless otherwise stated.

Converting imperial to metric

With the use of the Internet and overseas recipes, there is a continuing need to understand the conversions between both measuring systems. Below is an estimation of the conversions, and whilst they are as accurate as feasible, there may be differences of a few grams between amounts.

Dry measures

Metric	Imperial
15 g	½ oz
30 g	1 oz
60 g	2 oz
90 g	3 oz
125 g	4 oz (¼ lb)
155 g	5 oz
185 g	6 oz
220 g	7 oz
250 g	8 oz (½ lb)
280 g	9 oz
315 g	10 oz
345 g	11 oz
375 g	12 oz (¾ lb)
410 g	13 oz
440 g	14 oz
470 g	15 oz
500 g	16 oz (1 lb)
750 g	24 oz (1½ lb)
1 kg	32 oz (2 lb)

Liquid measures

Metric	Imperial
30 ml	1 fluid oz
60 ml	2 fluid oz
100 ml	3 fluid oz
125 ml	4 fluid oz
150 ml	5 fluid oz (¼ Pint, 1 gill)
190 ml	6 fluid oz
250 ml	8 fluid oz
300 ml	10 fluid oz (½ pint)
500 ml	16 fluid oz
600 ml	20 fluid oz (1 pint)
1000 ml (1 litre)	1¾ pints

Oven temperatures

In recipe books, instead of giving a numerical amount or measure, some recipes use descriptive terms for oven temperatures. Below is an interpretation of these terms and conversions to Fahrenheit.

Description	Setting (°C – Celsius)	Setting (°F – Fahrenheit)
Plate warming	60	120
Keep warm	80	160
Cool	100	200
Very slow	120	250
Slow	150	275-300
Moderately slow	160	325
Moderate	180	350-375
Moderately hot	200	400
Hot	220	425-450
Very hot	240	475

Some ovens which are fan-forced will require you to refer to the manufacturer's handbook for cooking guidelines. Generally, a fan-forced oven uses slightly reduced temperatures compared to an oven which is not fan-forced. A fan-forced oven generally cooks slightly faster than a conventional oven.

KNIVES

The knife was one of the first utensils developed for food preparation. Past materials used were stones, wood, even bone – materials that could be sharpened – and are vastly different to the sophisticated steel available today. Knives are a cook's basic tool and care needs to be taken with their use, as they are often very sharp. They vary according to the structure of the handle and blade, with the chef's knife the most common one used because of its versatility. You should be aware of the parts of a knife:

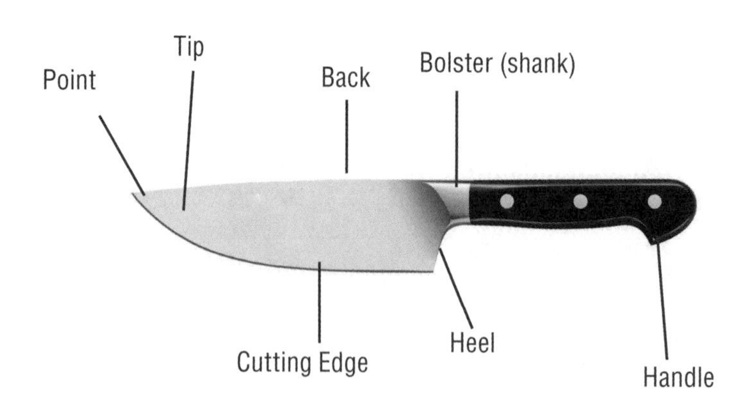

Types of knives

There are different types of knives for different jobs. These are detailed below:

- **Chef's or cook's knife:** has a blade between 20-25 cm long that is wide at the heel and tapers at the point. The handle is offset to allow the knuckles room so they do not bump against the chopping board.

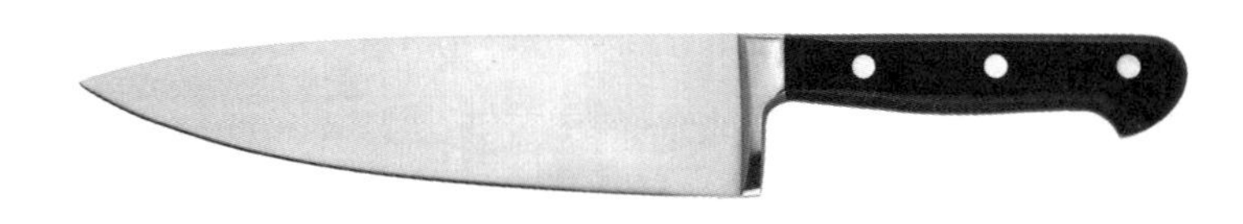

- **Boning knife:** has a thin pointed blade around 12 cm that is used for removing the flesh from the bones of raw meat and poultry.

- **Paring knife:** has a blade about 7 cm in length, all-purpose and useful in the finer cutting aspects of vegetable and fruit preparation. Can be known as a 'vegetable' knife.

- **Turning knife:** has a blade that is around 5 cm long and curved so that is useful for fine work like turning potatoes and garnishing/decorating other foods.

- **Serrated or 'bread' knife:** has a blade usually between 18-25 cm long with small cutting teeth that enable products to be cut without collapsing.

Sharpening knives

It is important keep knives sharp as they are safer than dull ones, since less pressure is required to cut food items and the blade is less likely to slip off the food. They are also more efficient and easier to use. Knives can be sharpened using whetstones (sharpening blocks) and a steel. Grinding a knife is only necessary if the knife has had prolonged or poor use and then only a professional should do it.

Using a stone

1. Coarser side upwards, place the stone securely on the bench with a damp, folded cloth under it to keep it from slipping.
2. Rub oil, detergent or even water as according to manufacturer's directions over the surface to act as a lubricant.

3. Hold the knife by the handle and rest the fingers of your free hand on the side of the blade, not too close to the cutting edge.
4. Place the knife at a slight 10° to 20° angle to the stone.
5. Move the knife along the length of the stone using long, even strokes and firmly pressing down on the blade. Press the heel of the blade down first, moving to the point then swap sides and repeat the process continually.
6. After sharpening run the knife down the side of the steel to remove any burrs.
7. When finished, rinse the blade under cold water to remove any metal filings, and dry the knife carefully.

Using a knife steel

A knife steel is used to hone the knife's edge – to get the point back on the edge of the knife. Use the following steps to practise the technique.

1. Hold the steel at a 45° angle with your thumb facing up – usually using your non-preferred hand.
2. With the knife in your preferred hand, hold the heel of it at 10° to 20° against the tip of the steel. Using a slicing motion, slide the knife down the length of the steel, moving from the heel to the tip.
3. Stroke the other side of the knife against the steel using the same motions. Usually only two or three strokes per side are needed.
4. Rinse and dry the knife carefully.

Storing knives

Because of their sharpness, knives need to be stored carefully. There are several ways this can be done safely.

- **Knife cases (like brief cases)** and **knife wraps or rolls**, which are usually made of leather or vinyl, are the best options for portable storage containers.
- **Toolboxes:** where each knife needs to be protected with its own cover when stored with other utensils.
- **Wooden storage blocks:** which are perhaps the most common way knives are stored in households.

Activity 7.5: Knives and related tools

Investigate the different types of knives used for different tasks in the kitchen. Complete the table below.

Type of knife/tool	Illustration	Purpose or use
Chef's or cook's knife		Used for slicing, dicing, cutting. The hard work or general purpose knife
Palette knife		
Parisienne scoop		
Carving knife		
Carving fork		
Slicing knife		
Meat cleaver		
Filleting knife		
Knife steel		

- **Magnetic knife racks:** are useful if space is limited but care needs to be taken when placing and retrieving them.

Activity 7.6: Knife cleaning safety

Personal injury from knives can often occur when the person is cleaning it. Describe the best method for cleaning a knife and detail two 'rules' you think necessary for handling knives during this process.

Knife technique

It is important to choose the proper knife for the intended task as wrong usage can lead to damaging the food product or, worse, injury to the user. The way you grip the knife is also very important as the proper technique will give maximum strength, accuracy and speed, reducing the risk of accidents occurring.

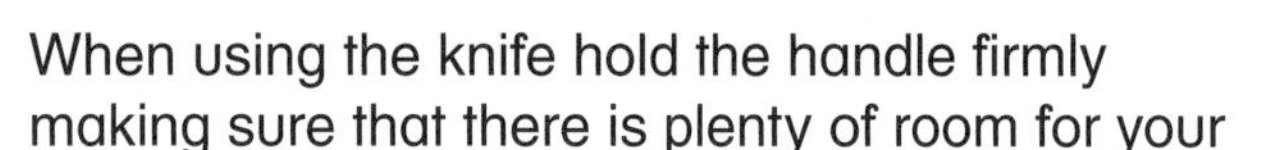

When using the knife hold the handle firmly making sure that there is plenty of room for your thumb and fingers to grip the handle so that when the heel of the knife touches the board, the fingers are not squashed. Make sure the grip is firm but relaxed since if too much pressure is used, you will tire more easily.

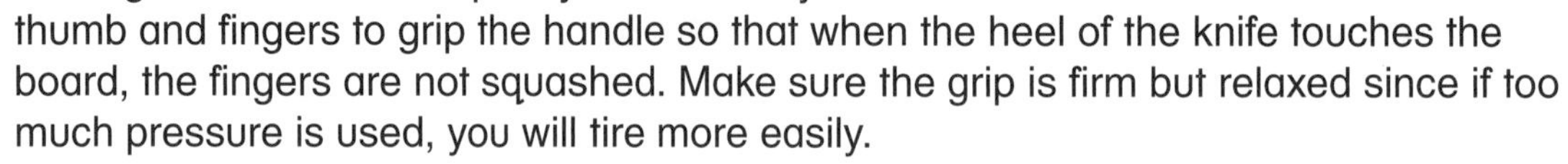

Correct knife grip

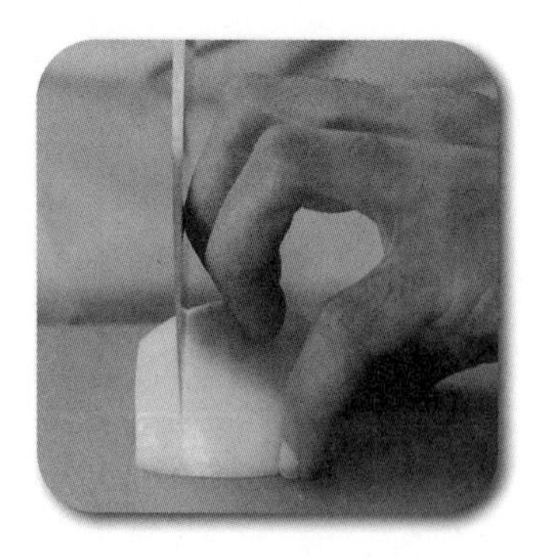

The food that needs to be cut should be held by the other hand with the fingers curved in a claw-like grip so that the knife blade is close to the knuckles. The fingertips are then protected and, with experience, the knuckles of the free hand can assist to guide the knife blade in the slicing process. In normal usage with general food items, the point of the blade needs to be on the board at all times. This increases speed and reduces the risk of accidents, as there is less room for the knife to slip or slide around. An even cutting rhythm is developed by ensuring that the blade of the knife rests squarely against the fingers and that the fingers move back as the cutting progresses.

It is important to practise the correct grip – it may feel awkward at the start but with practice should be easier and will make the cutting of food quicker and safer.

The cutting surface also needs to be either wooden or nylon which are yielding surfaces as hard surfaces like ceramic, stone or metal tend to damage the blade of the knife.

CUTTING FOOD

Using knife blade

When preparing food, to get the most benefit of the knife, you need to use different parts of the blade for different jobs.

The thin and narrow tip of the blade is best suited for fine and delicate work like for garnishing detail or cutting up small items similar to small vegetables such as onion.

The centre of the knife's blade is the most common part used in general work like chopping or slicing vegetables and meats.

The heel of the blade is used whenever cutting thick, dense items and a lot of force is needed. Examples of items include small bones, the roots of some Asian vegetables, for example bok choy, or other vegetables, for example a bunch of celery or leeks.

During food preparation – the cutting of food can be categorised as two methods – rough cutting and precision cutting.

Rough cutting

This refers to the practice of cutting foods coarsely when they are not needed for formal meal presentation. For example, vegetables are chopped roughly whenever they are needed to impart flavour to the particular dish that includes stews, braises, stocks and soups. Once the food has been cooked, the vegetables are then either pulverised like soups, or strained and discarded as in the case of stocks and some braises.

Mirepoix is a mixture of onions, carrots and celery used to flavour many stocks and stews, and is a typical example of a rough method of cutting.

Precision cutting

This refers to the cutting of food into specific shapes and sizes. The uniform sizes and shapes of the food can make the presentation of a dish special when used in salads, as accompaniments to main dishes or as a garnish. There are five specific terms for precision cuts that are usually used with fruit and vegetables though some meats lend themselves to these techniques.

- **Julienne:** makes long, thin strips (like matches) 3 mm x 3 mm x 40 mm. It is a common type of cut that can vary with different sizes. To julienne foods, top, tail and peel the pieces if necessary. Cut into lengths of 40 mm and square the edges off so there are no curves left. Slice into thin strips 3 mm thick, then stack the pieces into manageable heaps and slice lengthwise so 'matchsticks' are obtained.

- **Brunoise:** makes a very fine dice with 3 mm dimensions. Cut the same way as for julienne but then the match-like strips are cut into 3 mm lengths to form the tiny dice.

- **Jardinière:** makes batons 4 mm x 4 mm x 20 mm – usually used as a garnish. Cut the same way as for julienne but the pieces are cut into 20 mm lengths then 4 mm strips, then 4 mm lengths.

- **Macèdoine:** makes 8 mm dice. Usually used for fruit or vegetables such as either potato or fruit salads. Use the same techniques as described above but with 8 mm measurements instead.

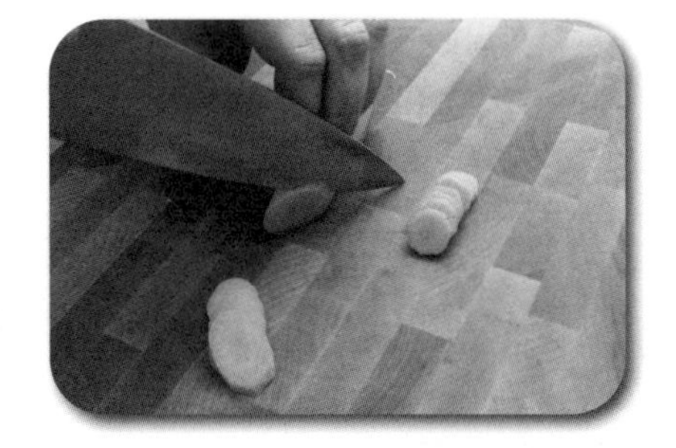

- **Paysanne:** are thin slices of food cut into 15 mm widths – crescents like celery, rounds like carrots or triangles and squares from other foods. For any shape food, simply slice 15 mm wide.

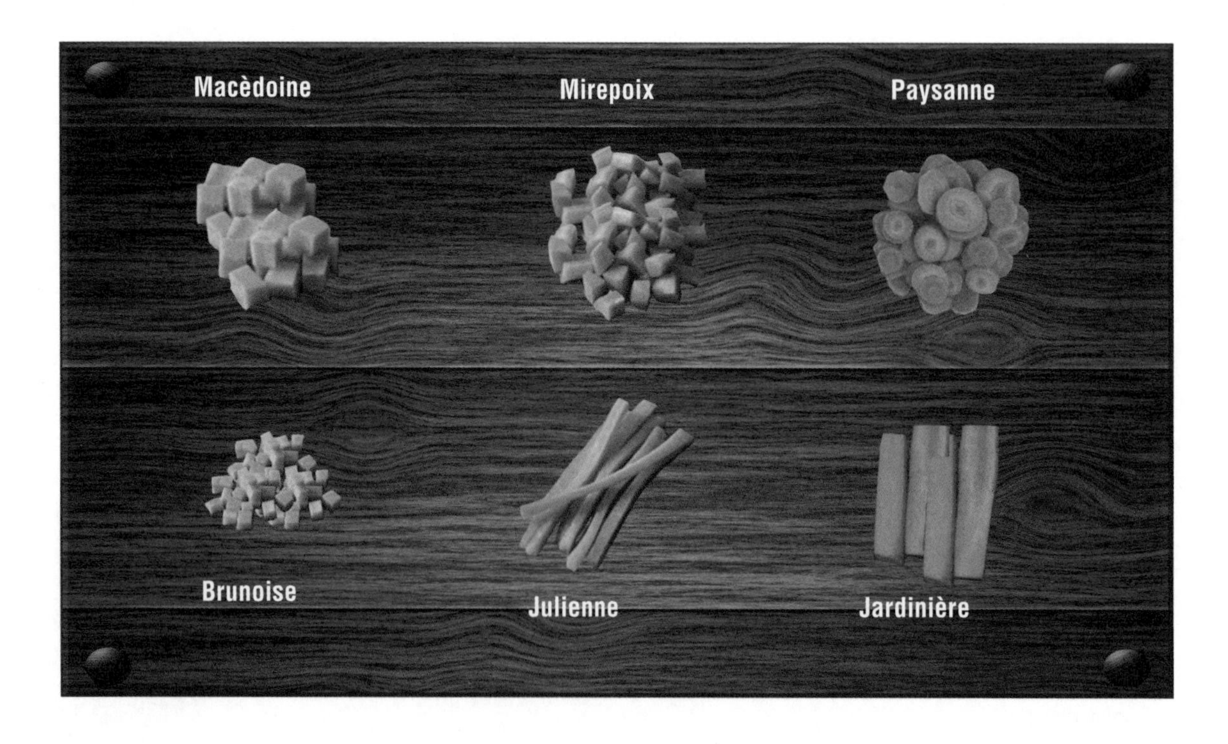

As well as the rough and precision cuts listed above there are other terms that describe the cutting of food. These are:

- **Chop:** refers to an uneven cutting method using short, sharp cuts with the final product being of a fine, medium or coarse cut. Applies to herbs.
- **Crush:** means squashing the food until it is bruised or reduced to fine particles often by using the side of a chef's knife blade. Mostly applies to garlic although some herbs can also be crushed.
- **Shred:** means to slice or grate into very thin strips as in the cabbage used in coleslaw and other leafy vegetables used in salads. Very fine shredding is known as 'chiffonade'.
- **Slice:** means using a sawing action to cut a thin, broad and flat piece of food – commonly onion rings or slices of bread.

Turning foods

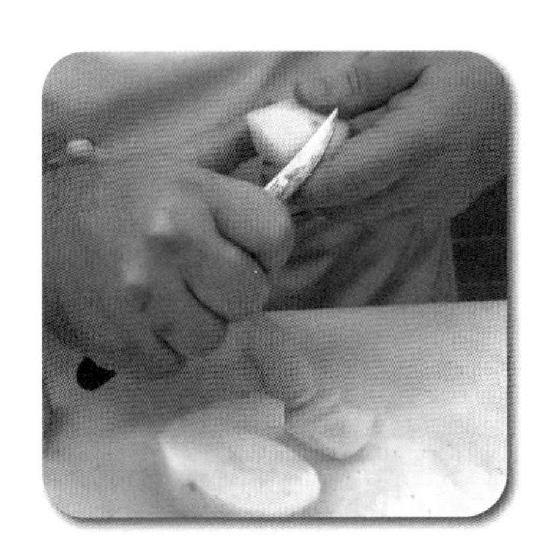

This refers to a technique where the food – usually root vegetables, are shaped into even-sized 'barrels' for the purpose of even cooking and pleasing visual appeal. A 'turning' knife is used because of its size and curved blade, but if that is not available, a paring knife will suffice. The food to be turned is to be cut to the required length, then the knife run down the outside evenly until there are six even, symmetrical sides.

Potatoes and carrots are useful for practising this technique on.

WASHING/CLEANING AND CUTTING CERTAIN FRUITS AND VEGETABLES

This is an important part of food preparation that, once done properly, will ease the cooking process in a busy kitchen. It involves the washing, cleaning, trimming, peeling, chopping, dicing and slicing of certain fruits and vegetables. You need to practise the techniques slowly at first to perfect them before concentrating on improving your speed so as to avoid accidents occurring. After the description of the various processes, you will find tables for 'Self-evaluation'. Please complete them as accurately as possible being a true critic, as through the self-evaluation process, you will be able to identify any weaknesses and be able to practise these, knowing the standard required.

Washing and cleaning

As most vegetables are grown either below or just above the ground it is important to clean them carefully to remove all traces of chemicals, pesticides, insects, dirt and fertilisers that may come with them. Use the following guidelines to ensure the vegetables are cleaned properly.

- Wash all vegetables in cold water – avoid soaking them as this could lead to the leeching of water-soluble vitamins. Some leafy vegetables may need to be washed several times to remove dirt – especially lettuce, spinach, bok choy and silverbeet. Make sure that they are drained well with excess moisture removed.
- Pay special attention to checking broccoli and cauliflower for insects – green caterpillars love to hide and can be difficult to find in the broccoli!
- Unwashed root vegetables that need to keep their skin on for cooking need to be cleaned with a stiff brush under running water.
- Leeks can be difficult to clean, so slice them lengthwise ¾ the way to the stem, then turn them sideways and do another lengthwise cut the same to create a 'cross' and then separate the layers and wash under running water.
- The stalks of celery need careful attention to remove all traces of dirt – use a brush or run your fingers down the stalk under running water to clean them thoroughly.

BASIC FOOD PREPARATION TECHNIQUES

This section will introduce you to a few of the basic concepts of food preparation, with emphasis on those preparation processes that are often the building blocks to further more elaborate dishes.

Peeling and trimming

After the cleaning process is completed, certain vegetables need to be peeled and trimmed using utensils like a paring knife, chef's knife and, of course, a peeler. These are used to remove roots, stems, strings, pith, eyes, skins and any blemishes that may be there. Some of the trimmings can be used and, instead of being discarded, should be added to stocks, sauces, mirepoix and purees. A good kitchen will always have one of these 'on the go' so that the trimmings are not wasted. The peelings that are the skins and tops and tails of vegetables usually have no use and can be thrown away. When peeling, only scrape the surface away as many nutrients lie just below it and need to be retained. Thin scrapings also reduce wastage and help keep costs down. Some foods can go brown if peeled and then exposed to the air. Fruits such as apple and pear and vegetables like eggplants need to have an acid like lemon juice poured over them or be immersed in a mixture of water and acid to stop this process, whereas potatoes just need to go into cold water until needed. Brine can also be used to stop the browning process.

Activity 7.7: Vegetable peelings

Some peels of vegetables are edible and can be used in the cooking process. How many can you think of?

Chopping parsley

1. Separate parsley from the stem.
2. Wash parsley in cold water.
3. Remove the stalks and finely chop the leaves using a chef's knife.
4. Place in a clean cloth like a piece of muslin, fold so the parsley is enclosed within and rinse under cold water.
5. Squeeze out excess moisture and place the dry and loose parsley in a bowl.

This process is done to make the parsley easier to sprinkle over dishes as a garnish – it should feel a bit 'paperish'. If the parsley holds too much moisture, it will stick together and be difficult to work with. This parsley can also be used in compound butters, sauces, duxelles and soups.

Washing, cleaning and cutting of fruit and vegetables	Self-evaluation	
Chopping parsley	✔	✘
• Separated from stem		
• Washed in cold water until dirt free		
• Excess water shaken off		
• Parsley is chopped finely		
• Parsley is thoroughly dried		

Peeling a tomato

This is the best method of peeling a tomato without damaging or removing too much of the flesh.

1. Wash a tomato and then cut a small cross into the bottom of it **(not the stem)** using the tip of a paring knife.
2. Remove the eye of the tomato – where the stem is.
3. Place a fork into the stem area, then plunge the tomato into rapidly boiling water for 10-15 seconds, turning the tomato frequently.
4. Immediately refresh it in iced water, and then remove the skin starting at the cross made in the bottom of the tomato. Please note – if the skin is already coming away from the tomato, it has been in the boiling water too long and has started to cook.

Washing, cleaning and cutting of fruit and vegetables	Self-evaluation	
Peeling a tomato	✔	✘
• All peel removed		
• Flesh not overcooked		

Mirepoix

This mixture of vegetables that usually contains carrots, onions and celery in equal quantities is used as a means of flavouring dishes – especially stocks and sauces. The vegetables are cut into rough dice in equal sizes, and sometimes herbs or a bit of bacon can be added for extra flavouring.

1. Wash the celery sticks, remove some of the leaves. Peel the onions leaving the roots attached.
2. Wash carrots, trim stems and bottoms and peel only if necessary (to remove blemishes or discolouration).
3. Roughly chop the vegetables into even sized pieces and mix together thoroughly. The size will depend on the use of the mirepoix – smaller sizes for smaller cooking times.

Washing, cleaning and cutting of fruit and vegetables	Self-evaluation	
Mirepoix	✔	✘
• Equal quantities of onions, celery and carrots		
• Vegetables washed and peeled, carrots topped and tailed		
• Cut to size suited for cooking time		

Slicing an onion

Sliced onion is used in braises, vegetable dishes, stir-fries, stews and some sauces.

1. Slice the top off the onion.
2. Peel the outer 'papery' layers off.
3. Cut the onion in half vertically through the stem to the root.
4. Leave the root and hold the rest of the onion together firmly.
5. Slice the onion finely by cutting even strokes from the stem to the root, angling the knife towards the centre as you go. Discard the root left over.

Washing, cleaning and cutting of fruit and vegetables	Self-evaluation	
Slicing onion	✔	✘
• Free of skin		
• Even thickness – approximately 3 mm		
• No root		
• Most onion used		

Dicing an onion

A diced onion is widely used in cooking and can be quite difficult to do due to the structured layers of the vegetable. You need to concentrate carefully with this technique because of the way the knife needs to be handled.

1. Peel onion and cut in half from stem to root. Slice the stem off but leave the root as this will hold the onion together.

2. Lay the cut surface of the onion on a cutting board and, using the tip of the knife, slice lengthways down the onion at 2 mm intervals not cutting through the stem.

3. Make 2-4 cuts at right angles to the previous cuts, again not slicing through the stem. The number of cuts will depend on the size of the onion. Because of the stem, you should be able to pick the onion up at this stage without it falling apart.

4. Parallel to the stem, slice at 2 mm intervals from top to bottom. The pieces should fall away from the onion in 2 mm small dice.

Washing, cleaning and cutting of fruit and vegetables	Self-evaluation	
Dicing onion	✔	✘
• Even 2 mm dice		
• Root base intact thoughout the cutting procedure		
• Sliced both vertically and horizontally before final dice		

Preparing citrus fruit

Fruits like oranges, lemons and to a lesser degree grapefruit are useful for garnishing and can be prepared into different shapes. The following details three different techniques. These tasks will be easier if done with a turning knife.

A. Peeled slices

1. Remove both ends of the fruit so that the flesh is just exposed.
2. Stand the fruit on a cutting board with one of the flat ends down, and slice the skin off using a slightly curved action, making sure all the pith is removed but not cutting into the flesh too much.
3. Turn the fruit sideways and cut slices 3 mm thick, removing any pips as you go.

Washing, cleaning and cutting of fruit and vegetables	Self-evaluation	
Peeled fruit slices	✔	✘
• Free of pith and peel		
• No pips present		
• Even width slices – approx 3 mm		

B. Wedges

1. Cut away both ends of the fruit without cutting into the flesh.
2. Stand the fruit on the cutting board and slice lengthways, then cut each half into four to six wedges depending on the size needed.
3. Remove any pips and cut off the strip of pith that is in the centre of each wedge leaving only flesh behind.

Washing, cleaning and cutting of fruit and vegetables	Self-evaluation	
Cutting citrus fruit wedge	✔	✘
• No pips		
• Uniform size		
• No pith in the centre of each wedge		

C. Segments

1. Remove both ends of the fruit so that the flesh is just exposed.
2. Stand the fruit on a cutting board with one of the flat ends down, and slice the peel off using a slightly curved action, making sure all the pith is removed but not cutting into the flesh too much.
3. Working over a basin or bowl, hold the fruit in one hand and remove each segment by slicing closely and carefully next to each membrane. Remove any pips, and place segment aside ready for use.
4. Squeeze excess moisture from the core left in your hand into the bowl.

Washing, cleaning and cutting of fruit and vegetables	Self-evaluation	
Segmenting citrus fruit	✔	✘
• No pith and peel in segments		
• No central membranes in segments		
• Segments intact and of uniform shape		

Peeling and crushing garlic

This strongly flavoured herb can affect other dishes with its aroma so should be one of the last tasks in the mise en place preparation.

1. Peel garlic clove by placing it on the cutting board and 'flattening' it firmly with the side of a chef's knife.
2. Once the peel is broken, remove the garlic clove.
3. Finely chop the garlic with a chef's knife then add salt to it with a ratio of a quarter part salt to one clove.
4. Using the flat of the knife blade, grind the mixture together until it forms a smooth paste.
5. Store end product in oil until needed for use.

Please remember to wash the cutting board immediately in hot soapy water to remove any traces of the garlic and its aroma.

Washing, cleaning and cutting of fruit and vegetables	Self-evaluation	
Peeling and crushing garlic	✔	✘
• Free of peel		
• Paste-like texture		
• Correct ratio of salt		
• Submerged in oil for storage		

BASIC MISE EN PLACE PREPARATIONS

There are quite a few basic preparations for mise en place that can be done beforehand and aid in the smooth flow of a food production area during the busy times. Follow the instructions carefully and complete the Self-evaluation charts as you progress.

Bouquet garni

This refers to a bunch of herbs and vegetables tied together to impart flavour to dishes like stews, soups, casseroles, stocks and sauces. The following describes a typical example of one. Please note that the size of the bouquet garni depends on the amount of stock it will be used in.

1. Cut two sticks of celery approx 10 cm and one piece of leek the same size.
2. Place side by side and then add a bay leaf, a sprig of thyme and four parsley stalks on top.
3. Tie together firmly using butcher's twine.

(The bundle is tied together so that it can be lifted out of the mixture at the end of the cooking time and easily discarded.)

Basic mise en place food preparations	Self-evaluation	
Bouquet garni	✔	✘
• Correct ratio of ingredients used		
• Correct proportion to the recipe requirement		
• Tied securely		

Marinades

Marinades are useful in cooking in that they are used to impart flavour, tenderise and preserve different foods. They vary from cooked ones to uncooked ones and from liquid to dry ones resembling a paste. A lot of different sorts of ingredients are used but usually there is a main ingredient like red or white wine, soy sauce and herbs and spices. A type of oil is also useful to help disperse the marinade over the food. Food that is marinating needs to be placed in glass, stainless steel or ceramic containers as the strong flavours and acid content of the marinade can affect certain plastic and some metal containers.

1. Assemble marinade ingredients and food to be marinated.
2. Mix marinade ingredients together well.
3. Pour over or coat food with the marinade evenly.
4. Cover and refrigerate. Baste frequently if necessary before cooking.

Basic mise en place food preparations	Self-evaluation	
Marinades	✔	✘
• Correct basic ingredients used		
• Item evenly coated or immersed		
• Marinated for correct amount of time		
• Refrigerated during the marinating process		

Compound butters

A compound butter is any butter with flavouring like herbs added to it. The herbs or other flavouring needs to be processed finely then mixed well with the butter. Compound butters can be whipped for aeration or for a smoother texture, and they can be piped for decorative purposes. They usually require at least one hour refrigeration before being able to be sliced for serving. The butters keep for a long time refrigerated or frozen as long as they are well wrapped, preferably in greaseproof paper. Types include garlic butter and parsley butter and these are mainly served with steaks.

Basic mise en place food preparations	Self-evaluation	
Compound butters	✔	✘
• Correct proportions of flavourings used		
• Portions easily without losing shape		
• Aerated to correct volume (almost white, light and fluffy)		
• Appropriately stored		

Clarified butter

Otherwise known as ghee, clarified butter is often used in Indian cooking. It is useful in cooking as it can be heated to a higher temperature than ordinary butter before it begins to burn, an excellent product for sautéing.

1. Melt butter very gently either in a saucepan on the stove top or in a bowl over a saucepan of boiling water (this would ensure that it does not burn).
2. When the butter has melted completely, use a ladle to skim off any scum that is on the surface and discard it.
3. Pour the rest of the butter into a strainer lined with muslin to separate the clarified butter from the whey which is not required and can be discarded.

Basic mise en place food preparations	Self-evaluation	
Clarified butter	✔	✘
• No impurities		
• Golden colour		
• Flavour is of pure butter		

Croutons

These are small pieces of fried bread used for garnishing or adding 'crunch' to soups and salads.

1. Remove crusts from sliced bread. Cut into 10 mm cubes.
2. Heat clarified butter in frying pan and fry croutons, tossing them in the pan so that they brown evenly.
3. Drain on absorbent paper once they reach a golden brown colour.

Basic mise en place food preparations	Self-evaluation	
Croutons	✔	✘
• Fried in clarified butter		
• Even golden brown in colour		
• Drained well so excess fat is removed		
• Crisp texture		
• Free from burnt crumbs		

Roux

A roux is a white mixture that is made from equal quantities of fat and flour that are blended and cooked together. For thicker mixtures, the proportion of flour is increased and although butter is the preferred fat to use, margarine or dripping can be substituted;although this affects the flavour. This mixture is the foundation for most sauces. Recipes can be found in Appendix A.

To make the roux firstly melt the butter, Remove from heat and add the flour blending thoroughly. Cook over a moderate heat to prevent the mixture scorching, stirring constantly to ensure the starch cells can develop to avoid a 'raw' taste.

A. White roux

The butter and flour are cooked lightly anywhere between 1-5 minutes depending on the quantity with a moderate heat, but is not allowed to colour. The mixture will develop into a crumbly texture during the cooking process but will come smooth again with stirring indicating that it is cooked. This is suitable for a white sauce, using milk.

B. Blond roux

The roux mixture is cooked in the same way as for the white roux but left on the heat until it becomes a light golden colour – usually between 6-7 minutes. This is used for veloute-based sauces where stock is used instead of milk.

C. Brown roux

This is used to thicken rich brown sauces or gravies and cooks the longest – about 8-15 minutes. The flour is slowly cooked until it reaches a rich hazelnut colour with a pleasant 'baked' smell. Because of the slow cooking process, part of the flour breaks down into dextrin which does not thicken as well as normal flour hence the sauce will not be as thick as the white or blond roux.

Once the required colour of the roux has been obtained, it should be cooled slightly before liquid is added, usually in the form of milk or stock. Slowly whisk the liquid into the roux then stir it constantly over heat until thickened.

Troubleshooting!

The roux can be challenging to make and can easily go lumpy. Read and remember the following:

- The roux mixture can separate – identified by fat floating on the top. This is an unrecoverable process and the mixture should be discarded.
- A grainy mixture is caused by cooking the roux too rapidly.
- Constant stirring is needed throughout the whole process to avoid lumps occurring.

Basic mise en place food preparations	Self-evaluation	
Roux	✔	✘
• Appropriate colour		
• Smooth texture – no lumps		
• Raw flavour not present		

Panada

This is a thick mixture of butter, flour and a liquid – usually milk or water – and it is used for thickening and binding various dishes like creams, soufflés, croquettes, stuffings and rissoles. Breadcrumbs combined with milk can also make a bread panada.

A. Flour panada

300 ml water
150 g flour
60 g butter

1. Bring water and butter to the boil.
2. Add flour away from the heat and mix well.

3. Return to the heat, stirring constantly until the mixture leaves the side of the pan.
4. Spread the panada onto a lined tray and brush with extra melted butter to prevent a skin developing on top of it. This should be a pale yellow colour and feel firm yet spongy to the touch.

B. Bread panada

600 ml milk
pinch salt
500 g fresh white breadcrumbs

1. Bring the milk to the boil and add salt, breadcrumbs mixing together thoroughly.
2. Remove from the heat and let cool for 2-3 minutes.
3. Return to the heat and constantly stir until the mixture is thick and leaves the side of the saucepan.
4. Spread the panada onto a lined tray and brush with extra melted butter to prevent a skin developing on top of it. Should be an ivory colour and have a fine crumb texture.

Basic mise en place food preparations	Self-evaluation	
Panada	✔	✘
• Appropriate colour		
• Correct textures		
• Smooth – free from lumps or fine crumb texture		

Tomato concasse

This refers to peeled and seeded tomatoes cut into 1 cm dice, a mixture used in many recipes such as pasta dishes, bruschetta and salads.

1. Peel the tomatoes as described previously in the chapter.
2. Cut in half horizontally (so you are **not** cutting the stem or bottom).
3. Remove the seeds by cutting the stem that holds them there and scraping them away with either your finger or a spoon.
4. Cut each half into 1 cm dice.

Basic mise en place food preparations	Self-evaluation	
Tomato concasse	✔	✘
• Flesh firm – no signs of cooking		
• Lack of seeds or skin present		
• Relatively even 1 cm dice		

Duxelles/'Mushroom hash'

This is a French term that means finely chopped mushrooms with a little onion and herbs, sautéed in butter until soft and most of the moisture has evaporated. It is used in vegetable and meat stuffing or can be enclosed in pastry and cooked until crisp.

Duxelles

20 g butter
500 g mushrooms
salt, white ground pepper
½ onion
1-2 sprigs of parsley
fresh breadcrumbs if needed

1. Slice whole mushrooms as thinly as possible, then pile them together and chop them finely.
2. Melt butter, add onions and sweat well.
3. Add mushrooms and cook gently until moisture has mostly evaporated away.
4. If too moist, add fresh breadcrumbs to absorb the extra liquid.
5. Add parsley and season to taste.

Basic mise en place food preparations	Self-evaluation	
Duxelles	✔	✘
• Greyish to dark in colour		
• Fine, crumb-like paste		
• Moist, spreadable consistency		
• Rich, deep mushroom flavour, suitably seasoned		

BATTERS AND COATINGS

Batters and coatings are used in cooking to protect and seal foods that are fried or deep-fried. This protection lets the food be cooked in the hot oil without falling to pieces while also sealing in the juices. There are various coatings that can be used and a proficient cook knows which ones are best suited for the different types of food. The coatings can be classified as either wet or dry, with batters commonly used as a wet type and breadcrumbs used as a dry sort. Batters and coating also contribute to the flavour and texture of the final product.

A. Batters

These are flour mixtures that can contain egg, yeast, beer or milk depending on its intended use. Both the yeast and beer are often used in the batters because of the light effect they produce and beer gives a particular tangy flavour to the end product. Any mixture that has been battered and fried needs to be immediately served as the product can become soft and soggy when left too long.

Beer Batter

(This will give a crisp coating to most deep-fried food)

2 c SR flour
½ c beer
1 c cold water

1. Whisk ingredients together until smooth.
2. Use as required immediately.

Variations: Substitute equal proportions of SR flour for chickpea flour or ground rice. Flavour if desired with ground spices, eg. cumin, pepper, turmeric.

Basic mise en place food preparations	Self-evaluation	
Batters	✔	✘
• Ivory colour		
• Smooth, free from lumps		
• Thick, creamy consistency		
• Yeasty beer flavour, suitably seasoned		

Besides the beer batter, there are other types of batters including ones used for tempura, crepes and pikelets. It is useful to practise making these as the process is very similar. Check out some of the recipes at the end of this book.

Activity 7.8: Batters

Batters can be used on a range of foods for deep-frying. List all the foods you think would be suited for this purpose.

B. Coatings

This is a dry covering for food made from dried breadcrumbs, fresh white breadcrumbs, nuts, wheatgerm or even breakfast cereals. Any foods requiring crumbing are initially prepared by trimming and thoroughly drying the item to remove any excess moisture. This ensures that it can be coated evenly with flour, the next step, before being dipped in egg and finally being covered with the crumbs. Once the process is complete, crumbed foods should be laid out separately in a single layer to ensure that they do not stick and refrigerated, even frozen, until ready to be cooked. Crumbed food products can also be refrigerated once cooked.

Crumbing process

1. Prepare food to be crumbed by trimming and removing excess moisture from the food product.
2. Dip the product into the flour, gently shaking off any excess.
3. Dip the product into the egg wash making sure it is completely coated with egg. Allow excess egg to drain off to allow an even coating of crumbs.
4. Place product into the crumbs and gently press them on. Gently shake off any excess after ensuring the product is thoroughly coated.
5. Place on single level on tray that is lightly scattered with extra crumbs.

Warning!

This can be an extremely messy process, but experienced cooks working by themselves know to keep one hand for the dry steps and the other hand for the wet one. That is, right-handed cooks use their right hands for the flour and crumbs and the left hand for the egg wash step, working from the left to the right. Left-handed cooks do the process oppositely.

Suitable foods for crumbing include fish, shellfish, cuts of meat like schnitzels, vegetables and even ice cream.

Fresh breadcrumbs

1. Remove crusts from the bread.
2. Cut into cubes roughly.
3. Use a food processor to finely crumb the bread.
4. Use immediately or store in the freezer.

Basic mise en place food preparations	Self-evaluation	
Fresh breadcrumbs	✔	✘
• No evidence of crusts		
• Fine-grained crumbs, not dried out		
• White colour		

Dried breadcrumbs

1. Remove crusts from the bread.
2. Cut into cubes fairly evenly.
3. Place on baking tray evenly and bake in oven at 180°C until light brown in colour.
4. Allow to cool and then process in a food processor until mixture is finely crumbed or crush with a rolling pin.
5. Use when needed, store in a cool, dry place.

Basic mise en place food preparations	Self-evaluation	
Dried breadcrumbs	✔	✘
• Light brown in colour		
• Fine crumbs		
• Crisp		

Useful Websites

www.youtube.com.au: Contains videos of knife skills and food.

CHAPTER 8
Methods of Cookery

Key Concepts

- Defining the cooking techniques
- Effects of processing techniques on the properties of food
- Dry cookery methods
  - Baking
  - Roasting
  - Frying
  - Grilling/barbecuing
- Wet cookery methods
  - Boiling and simmering
  - Poaching
  - Steaming
  - Stewing
  - Casseroling
  - Braising
- Microwave cookery

DEFINING THE COOKING TECHNIQUES

The actual cooking of food, that is, the application of heat to a food, comes with its own set of terminology, which you need to understand if you are to successfully interpret recipes and apply these terms to different foods.

Let's look at what happens when heat is applied to foods. The transfer of heat in cooking is in one of these forms: conduction, convection and radiation.

- **Conduction** is the transfer of heat through solid materials. When two objects of different temperatures are placed into contact, heat will flow from the object with the higher temperature to the one with the lower temperature.

  A flow-on effect occurs when molecules in the closest position or in direct contact with the higher heat source vibrate causing others nearby in turn to vibrate as well. An example is deep-frying a battered fillet of fish. The outside batter cooks quickly with the fish cooking next.

- **Convection** is when heat is transferred using the movement of liquids or gases as the transfer medium. When gases or liquids are heated their molecules expand. The hot molecules rise and in their place cooler molecules move. These are then heated and the process repeats, creating a current. An example is boiling liquids on a stove or in an oven where roasting or baking is taking place.

- **Radiation** is when energy is transferred by radiation in the form of rays. A heat source radiates waves that travel in straight lines towards food. When it reaches the food, it is absorbed, cooking quickly from the outside through to the middle. An example is a slice of bread in a toaster.

Different types of cooking methods utilise the different principles of heat transfer as outlined below. Usually a cook will decide what characteristics they want in a final food product and will utilise the appropriate cookery method to achieve that.

Did you know ...

In preparing and cooking food, such as a sponge cake, heat may be transferred by all three methods:

- convection - heat circulating in the air of the oven
- conduction - heat transferred from the cake pan to the cake mixture
- radiation - heat reflected off the oven walls

Effects of processing techniques on the properties of food

Different types of cooking methods utilise the different principles of heat transfer as outlined below. Usually a cook will decide what characteristics they want in a final food product and will utilise the appropriate cookery method to achieve that. Final characteristics include:

- **appearance:** eg. shape, colour, size, volume, viscosity
- **texture:** eg. hard, crisp, soft, malleable,
- **flavour:** eg. bitter, sour, sweet, salty, acidic, mild, fiery, spicy, tart
- **aroma:** eg. acrid, fresh, fragrant, spicy, savoury, sharp, sweet
- **palatability:** agreeable to the palate or taste
- **nutrition:** effect on the nutrients
- **quality and food safety:** food hygiene
- **digestibility:** ease of consumption
- **temperature:** suited to the food and occasion.

These will be outlined for each processing technique in the section to follow.

TABLE 8.1: Heat transfer in cooking

Style of cooking	Heat transfer	Elaboration
1. Dry cooking methods		
Grilling, Barbecuing	Radiation	Rays reach food directly
Baking	Convection	Hot air movement/currents
Roasting	Convection	Hot air movement/currents
Frying – deep	Conduction	Transfer of heat through oil
2. Wet cooking methods		
Boiling	Conduction, convection	Transfer of heat through saucepan, followed by hot liquid currents
Poaching	Conduction, convection	Transfer of heat through saucepan, followed by hot liquid currents
Stewing	Conduction, convection	Transfer of heat through saucepan, followed by hot liquid currents
Steaming	Conduction, convection	Transfer of heat through saucepan, followed by hot air currents
Braising	Conduction, convection	Transfer of heat initially through frypan, then through added liquids
3. Other methods		
Microwaving	Radiation	Electromagnetic waves cause water molecules to vibrate resulting in the formation of heat energy

Activity 8.1: Terminology

Find out what the following terms mean and give examples where you can.

- Bake
- Barbecue
- Brown
- Blanch
- Deglaze
- Grill
- Microwave
- Parboil
- Poach
- Roast
- Sauté
- Sear
- Simmer

DRY COOKING METHODS (OR DRY HEAT COOKERY)

Dry cooking methods employ very little use of moisture and are tailored to foods that are tender and/or smaller in size to cook. For example, grilling meats, baking biscuits or roasting vegetables are all examples of dry methods of cookery.

A. Baking

Description

Baking is a method of cookery often thought of as the most difficult to control. It has a lot of variables that need to be considered and some factors that are beyond control such as atmospheric pressure or humidity! Baking involves cooking foods using the dry heat created by an oven. It is a process often related to cereal products where the hot air circulates around the food to provide a dry, crisp result. The baking food does not come into contact with a cooking liquid yet in some cases a water bath may be used to prevent the intensity of the dry heat spoiling the cooking food, as in egg custards. Baking temperatures vary from around 100°C to 210°C.

Effects on properties of food

A main effect of baking of the properties of food is **dextrinisation** where heat makes any starch present turn golden brown. Food becomes crisp or firmer, losing water or liquid content. There is some nutrient loss depending on the temperature utilised. Protein will denature at high temperatures.

Equipment

Ovens are the main piece of equipment used along with baking containers or trays. Container materials are generally of a metal construction to provide effective conduction of heat to the food product. Baking dishes, cake tins, flan tins, muffin trays and baking trays are typical examples of equipment used in baking.

Pointers (addressing quality control):

- Smaller or tender foods are suited to this method.
- Check oven facilities such as fan-forced or source of heat. This will affect where you place your foods for baking in the oven.
- Some ovens have 'hot spots' where the food may burn while others are 'slower' or 'faster' requiring foods to be cooked at differing recommended temperatures. Become familiar with your oven and its peculiarities over a period of time.

For example, some conventional ovens are hotter at the top section compared to the lower levels of the oven.

- Be careful of food proportions as baking requires careful measuring to ensure a quality final product.

Safety

Baking involves heat and movement or the carrying of hot equipment and/or foods. Therefore attention needs to be taken to avoid burns, trips or falls when carrying hot objects. Correct ergonomic lifting processes need to be employed when working with a low oven. Care needs to be taken when lighting ovens that utilise gas. You also need to ensure the oven is off and cold when cleaning.

Food examples

Foods best suited to baking include:

- breads, scones
- cakes
- muffins
- biscuits and baked slices
- pies, quiches, flans
- custards (generally utilising a water bath)
- whole fish
- dishes such as frittata, lasagne, pasta bakes.

Recipes for baking

Chocolate Chip Cookies
(Makes 15)

125 g butter
½ c white sugar
½ c brown sugar
1 tsp vanilla essence
1 egg
1½ c SR flour
100 g chocolate chips

1. Preheat oven to 180°C. Lightly grease a baking tray.
2. Cream together the butter, sugars and vanilla. Add lightly beaten egg.
3. Mix in sifted flour.
4. Add chocolate chips and mix evenly.
5. Shape teaspoonfuls of the mixture into small balls and place on trays allowing room for spreading.
6. Bake for 10-12 minutes.
7. Remove from trays and place on cooling racks.

Choux Pastry
(Makes 24-30 puffs)

1 c plain flour	125 g butter
1 c water	3 eggs

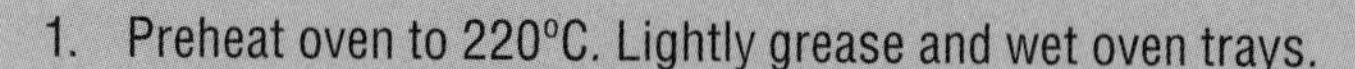

1. Preheat oven to 220°C. Lightly grease and wet oven trays.
2. Place butter and water in saucepan and stir while heating until butter is melted. Bring to the boil.
3. Remove from heat immediately and add flour quickly all at once. Stir with a wooden spoon until the mixture forms a smooth ball.
4. Allow to cool slightly.
5. Add eggs once at a time, beating thoroughly after each addition. Mixture needs to be smooth and glossy.
6. Cool completely before using. Mixture can be refrigerated until required.
7. Place spoonfuls of mixture onto trays allowing room for spreading.
8. Bake for 15 minutes then reduce heat to 180°C. Bake until golden brown and no beads of moisture are evident.
9. Place on cake cooler. To ensure a crisper puff, slightly split puff and remove uncooked mixture from the inside with a knife or spoon.
10. Puffs can be filled with cream or a confectioner's custard (see pastry cream recipe).

Simple Margherita Pizza
(Between 2)

100 g SR Flour (¾ c)	150 g pizza cheese
Pinch salt	½ c canned tomatoes, drained and chopped
30 g margarine	2 Tbsp fresh basil to taste
1 Tbsp cold water	salt, pepper to taste

1. Preheat oven to 220°C.
2. Combine half the flour with the salt, margarine and water. Mix with a fork. When creamy, add the rest of the flour and mix to a stiff dough.
3. Roll dough out into a circle (about as thick as a slice of bread). Put on a greased baking tray.
4. Spread chopped tomatoes evenly over the circle; then the cheese, then sprinkle over the herbs and season with salt and pepper.
5. Bake for 25 minutes in the oven. Slice and serve.

Did you know ...

Pizza first originated in the early 1700s in Naples, Italy, and the most popular topping in pizza in Japan is squid!

Sponge Cakes
(Makes 2 x 20 cm cakes)

4 eggs (60 g)
¾ c caster sugar
$\frac{1}{3}$ c plain flour
$\frac{1}{3}$ c SR flour
$\frac{1}{3}$ c cornflour

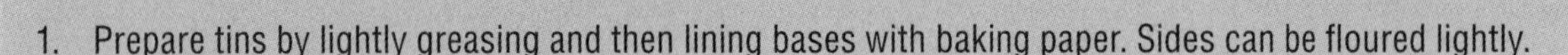

1. Prepare tins by lightly greasing and then lining bases with baking paper. Sides can be floured lightly.
2. Preheat oven to 170°C.
3. Beat eggs in large bowl until thick and creamy.
4. Gradually add in sugar 1 T at a time, beating well in between additions. Sugar needs to be dissolved after each addition – test by rubbing a small amount of mixture between fingers to test if smooth.
5. Sift all flours together 3-5 times to aerate mixture.
6. Fold the flours into the egg carefully and slowly. Ensure all flour is folded into the mixture evenly.
7. Pour equally into prepared tins and bake in oven for about 20 minutes or until sponge recovers when pressed lightly with finger.
8. Remove from tins and cool on wire racks.
9. When cold, fill and ice as desired.

B. Roasting

Description

Roasting is very similar to baking in that it involves an oven to circulate dry heat around a food. However, in roasting the foods are generally larger in size and are usually basted during cooking. During the roasting process the dry heat browns or sears the outside of the food and the food's interior is softened by heating its natural moisture or water content.

Mirepoix is a mixture of onions, carrots and celery used to flavour many stocks and stews, and is a typical example of a rough method of cutting.

Effects on properties of food

These are similar to baking. Where fat or oil is used there may be increased fat intake; fat soluble vitamins may be lost if fat is drained away.

How to roast meat

1. Preheat oven to 220°C. Check shelf positions.
2. Prepare a baking dish by either inserting a roasting rack (or trivet) or by placing a mirepoix of vegetables into the bottom of the baking dish.

3. Meat should be at room temperature. Weigh the meat to ascertain the cooking time required. Refer to the pointers below.
4. Ensure the roasting piece maintains a correct, even shape by tying with string or trussing as in poultry.
5. Place the meat in the roasting dish and season as required.
6. Cook in the oven for 15 minutes then reduce the temperature to 180°C for the remainder of the cooking time. This sears the outside of the meat to help prevent loss of moisture.

7. Baste throughout the cooking process so that the food remains moist.
8. Test for doneness by either piercing the outside flesh to check the colour of the juices or by using a meat thermometer inserted into the flesh.
9. Allow to rest for 15 minutes, covered, before carving.

Equipment

An oven and baking dishes are required for roasting. Sometimes a spit-roast is employed for whole carcasses such as pork or lamb.

Pointers (addressing quality control):

- **Roasting times:** (Note this is a guide only.) Times are dependent on the type of meat, whether bone is present and so on. Ask your butcher for more specific advice when purchasing. Remember that meats continue to cook for a while during the resting period.
- **Beef:** 210°C for first 10 minutes, reduce to 180°C for remainder.
  - Rare: 20 min/kg
  - Medium: 35 min/kg
  - Well done: 50 min/kg
- **Lamb:** 190°C, 35-40 min/kg
- **Pork:** 220°C, reduce to 190°C after 10 minutes, 50 min/kg
- **Chicken:** 190°C, 40 min/kg

Safety

As per baking, roasting involves heat and movement or the carrying of hot equipment and/or foods. Therefore attention needs to be taken to avoid burns, trips or falls when carrying hot objects. Correct ergonomic lifting processes need to be employed when working with a low oven. Care needs to be taken when lighting ovens that utilise gas. You also need to ensure the oven is off and cold when cleaning.

When roasting meats, sometimes a lot of hot fat is produced, so be especially careful when removing them from the oven that this doesn't spit or spill onto you. Also, some baking dishes are quite heavy so need to be carefully managed.

Activity 8.2: Roasting

1. Find out what accompaniments traditionally are served with the roast meats listed in the pointers. List them.
2. Why are foods basted and what mixture can they be basted with? Do you need any special equipment for this?
3. Describe a rotisserie. In what situations is it used?

Recipes for roasting

Roasted Lemon Chicken with Mediterranean Vegetables
(Serves 2)

Chicken:

2 chicken thighs, bone in
1½ Tbsp lemon juice
½ tsp lemon zest
1 tsp garlic
1 tsp thyme
2 tsp parsley, chopped
2 tsp oil
¼ tsp pepper
¼ tsp salt

1. In a shallow glass dish mix together the lemon juice, zest, garlic, thyme, parsley, oil and pepper. Add the chicken and turn until coated in marinade. Cover and refrigerate overnight.
2. Preheat oven to 200^0 C. Remove the chicken from the marinade and lightly season with salt. Place in a roasting tin ensuring dish is not overcrowded. Roast for 40 mins, or until cooked through. Serve on plate and drizzle with juices from cooking pan.

Vegetables:

2 t oil
2 small onions
1 small zucchini, quartered
1 small eggplant, quartered
½ red capsicum, sliced thickly
2 medium-sized potatoes, quartered

1. Heat oil in baking dish; add onion, zucchini, eggplant, capsicum and potato. Cook stirring, until vegetables are browned lightly.
2. Place vegetables in hot oven and cook uncovered for 30 minutes or until tender and browned.
3. Serve with chicken.

Roasted Beetroot, Feta and Pine Nut Salad
(Serves 8)

3 small beetroot
2 Tbsp olive oil
500 g sweet potato
200 g rocket leaves
150 g pine nuts
100 g feta cheese, cubed

Dressing:
3 Tbsp good olive oil
2 Tbsp lemon juice

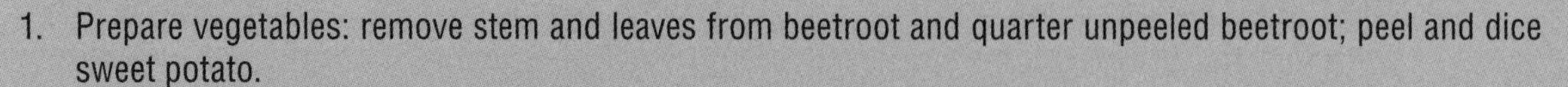

1. Prepare vegetables: remove stem and leaves from beetroot and quarter unpeeled beetroot; peel and dice sweet potato.
2. Toss beetroot in half the oil and place on lined baking tray. Bake in moderately hot oven for 15 minutes.
3. Toss sweet potato in remaining oil, place on baking tray with beetroot and continue to bake for 30 minutes or until vegetables are tender.
4. When beetroot is cool enough to handle, peel.
5. Toss rocket, beetroot, sweet potato, feta and pine nuts together in a large serving bowl.
6. Combine dressing ingredients in a jar and gently fold through the salad.

C. Frying

Frying is a quick method of cooking suited to small pieces of food. The high temperatures of the cooking medium (fat or oil) allow the foods to cook quickly creating a crisp, browning effect. There are different types of frying that are outlined below.

Effects on properties of food (frying)

Frying is a quick method of cooking generally producing a crisp, brown product. Short cooking time decreases negative effects of heat on nutrients. Fat is a pleasing flavor to most people (when hot) so frying increases a food's taste appeal. Foods are not colourful (except when stir-frying) – but are simply brown.

1. Dry-frying

Description

This is a method whereby the food is cooked on a hot surface without the addition of a cooking medium. The food itself generally contains enough fat to cook in itself, for example, bacon or sausages.

How to dry-fry

1. Prepare foods by trimming excess fats and cutting to size.
2. Preheat hot plate and spray with a little oil if required to prevent sticking.
3. Using tongs, place food on the hot plate and turn once sealed.
4. Once cooked remove and drain off excess fat.

Note: Fat may have to be drained off during the cooking process. To do this remove the foods, pour off fat quickly and then replace the foods to finish cooking.

Equipment

A hot plate such as a grill, barbecue or frypan are suited to dry-frying. A small spray of oil may be useful to prevent food initially sticking on these surfaces. A sandwich press is an electrical appliance that also uses the principles of dry-frying.

Pointers (addressing quality control):

- Ensure a high heat to melt the fat within the foods.
- Do not pierce foods such as sausages.
- Turn rounded foods frequently to create even browning.
- Drain excess fat where possible during the cooking process.

Safety

Frying foods needs hot surfaces and hot oil/fat. Be careful of spitting fats and take care to prevent burns. Hot fat/oil can catch fire – extinguish the source of oxygen by covering the flames with a lid and turn the appliance off. Never leave frying foods unattended.

Food examples

Foods best suited to frying include:

- sausages
- bacon
- hamburgers
- rissoles
- pancakes.

Recipes for dry-frying

Pancakes or Crepes

2 c plain flour
60 g butter
pinch salt
1½ c milk
3 eggs

1. Sift flour into basin and make a well.
2. Warm together the butter, salt and milk until butter has melted.
3. Break eggs into the well and work in some flour.
4. Add warm milk mixture gradually while whisking until smooth.
5. Refrigerate 2 hours before cooking. Consistency should be like thin cream – if not, adjust with more milk.
6. Heat frying pan and wipe with a piece of buttered paper.
7. Pour in some mixture and swirl to reach the edges of the pan.
8. Cook until edges lift slightly and turn with fingers or with a palette knife.
9. Cook other side then lift onto plate and use accordingly.

Pikelets
(Makes 15)

1 egg
3 Tbsp sugar
½ c milk
1 c SR Flour
2 tsp butter or 1 Tbsp cream

1. Place egg, sugar and half of the milk in a bowl.
2. Whisk well with a rotary or balloon whisk.
3. Sift flour and add to mixture.
4. Add milk until batter is the consistency of thick cream.
5. Add melted butter or cream. Beat until smooth.
6. Heat a lightly greased frypan or griddle.
7. Spoon batter into pan and turn when surface bubbles.
8. Remove and place in a towel to cool.

Activity 8.3: Frying

Take a range of sausages from a variety of providers and cook each separately for the same amount of time at the same heat. It may assist you in your results if you weigh each beforehand. Measure the amount of fat from each. Create a comparison of each using factors such as appearance, taste, fat released, texture etc. Then answer the questions 'what makes a good sausage?' or 'what do you look for in a quality sausage?'

2. Stir-frying

Description

This method employs high heat with minimal oil and the quick movement of foods in a wok or frypan. It is a healthy alternative compared to other frying methods and usually involves vegetables and small amounts of lean meat.

How to stir-fry

1. Prepare foods using mis en place techniques.
2. Keep foods with similar cooking times together.

3. If marinating ingredients, complete this beforehand.
4. Heat wok or cooking vessel. Add oil/s.
5. Cook meats in batches and then cook other foods as required by the recipe.
6. Stir-fries are usually served with rice or noodles.

Equipment

Traditional methods use a wok; however, if this is not available, a heavy based frypan can also be used. Metal ladles or spatulas are used for tossing the food while cooking; however, if using a coated wok or frypan use suitable plastic coated utensils to prevent scratching the surface of the pan.

Pointers (addressing quality control):

- Cook harder foods before softer, for example carrot before bean shoots.
- Try to keep pieces of food roughly the same size for even cooking.
- The wok must be hot before adding oil and foods.
- Cook meats in small batches of around 100 g to maintain the wok's heat and prevent the meat from leaching liquids and becoming tough.
- Season woks regularly to maintain their surface.

Safety

Be careful of burns and try to avoid lifting heavy frypans/woks full of food. Also, because of the surface area and small amounts of oil, the wok can easily catch alight so watch out for this hazard. Keep a lid handy to put on the wok if a fire occurs.

Food examples

Foods best suited to stir-frying include:

- vegetables
- fried rice
- meats combined with vegetables and/or noodles
- Asian dishes.

Activity 8.4: Stir-frying

1. Describe your favourite stir-fry and list the main ingredients. What makes an interesting stir-fry and what are some of the more unusual ingredients you have added to a stir-fry that you have made?

2. What is meant by 'seasoning' a wok? Describe the process you have to go through to get this done.

Recipes for stir-frying

Stir-Fry Chicken
(Serves 2)

250 g chicken thigh fillets
1 carrot, julienne
1 celery stick, julienne
1 small red capsicum, julienne
1 Tbsp oil
1 clove garlic, crushed
1 tsp red chilli, finely diced
½ tsp ginger, freshly grated
1 small onion, sliced
2 tsp soy sauce
1 Tbsp oyster sauce
1 tsp sesame oil
1 tsp brown sugar

1. Prepare vegetables. Set aside.
2. Remove fat from chicken and cut into thin strips.
3. Heat oil in wok over a high heat.
4. Add garlic, chilli, ginger and onion. Stir-fry for 2 minutes or until onion is soft.
5. Stir in chicken and cook until chicken is browned.
6. Add carrot, celery and capsicum and stir-fry until celery is just tender.
7. Combine sauces, sesame oil and sugar and stir into frypan.
8. Serve immediately.

3. Shallow-frying

Description

In this method enough oil is used to cover the base of the cooking vessel – typically a frypan – and the food is turned to allow for even browning.

To shallow-fry:

1. Prepare foods by trimming, cutting and coating if required.
2. Add and heat oil in the pan until almost reaching smoke point.
3. Carefully place in foods to be fried leaving space between items to turn them. Once golden, turn with tongs, being careful not to pierce the foods or damage them.
4. Once cooked drain on absorbent paper.

Equipment

A frypan is the main piece of equipment used for shallow-frying. Slotted spoons and tongs are some of the utensils used for this method.

Pointers (addressing quality control):

- Use absorbent paper to drain cooked foods.
- Use fresh oil and don't overheat. Oil which is smoking is too hot and will burn the outside of the food, not cooking the inside properly.
- Keep pieces of food the same size when cooking in batches.
- Avoid adding too many pieces of food at once as this lowers the temperature of the cooking medium causing it to become greasy.
- Remove loose foods or crumbs at the end of each cooking batch to prevent the oil from being tainted by burnt remains.
- Remove excess water from foods to be shallow-fried to prevent oil from spitting.
- Do not overcrowd the frying pan, allowing room for foods to be turned easily.
- Equal amounts of oil and butter (preferably clarified) add a nice flavour to foods. Be careful not to overheat as this combination requires a lower heating temperature than straight oil.
- Do not use butter on its own as a frying medium as butter burns very easily.

Safety

The safety issues are as per dry-frying. Also, be aware of mixing water with oil as this is a dangerous combination. Therefore, dry foods as well as you can before placing them in hot oil. Supervise the cooking at all times.

Food examples

Foods best suited to shallow-frying include:

- crumbed cutlets
- wiener schnitzel
- crumbed fish
- vegetable or fruit fritters.

Recipes for shallow-frying

Rice and Zucchini Patties
(Serves 4)

1½ c cooked brown rice
300 g zucchini, grated
1 onion, chopped
2 Tbsp fresh parsley, chopped
2 eggs, beaten
1 c stale breadcrumbs
130 g can creamed corn
¼ c polenta
60 ml olive oil

1. Prepare vegetables.
2. Combine rice, zucchini, onion, parsley, egg, breadcrumbs and corn in a large bowl and mix thoroughly.
3. Divide mixture into 12 and shape into patties. Coat lightly in polenta. Set aside on chopping board.
4. Heat oil in frying pan.
5. Cook patties in batches until well browned and heated through. Drain on absorbent paper and keep warm.
6. Serve either warm or cold.

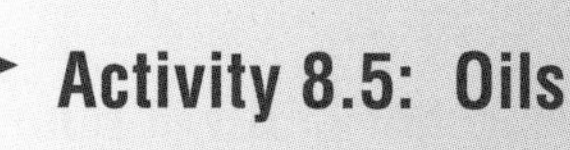

Activity 8.5: Oils

Investigate the types of oil available in supermarkets. Compare them in terms of type, purchasing, cost, use, flavour and interesting facts. To test flavour, use fresh bread such as ciabatta and dip it in each cold oil and taste. Which ones do you like and what did you find out from the activity?

4. Deep-frying

Description

This method employs a large amount of oil in a heavy based saucepan or deep fryer. Foods are coated to protect them (sometimes adding flavour) and the cooking time is quite quick.

How to deep-fry

Use the same method as for shallow-frying, except food generally does not have to be turned in the cooking process.

Equipment

It is necessary to use a deep vessel, namely a deep fryer, due to the expansion of the oil in the heating and cooking process. Deep fryers are generally electric so that the temperature can be thermostatically controlled to prevent overheating issues.

Pointers (addressing quality control):

- Ensure the fryer is no more than two-thirds full of oil, including the allowance for a basket if used.
- The frying temperature is usually 180°C-190°C and can be checked if ready by dropping a cube of bread into the heated oil. If the bread cube turns golden it is ready. If the oil is smoking it is too hot.
- Be careful to lower foods slowly into the hot oil, use a basket if able to and keep hands and body well clear.
- Some foods need to be coated to help protect them from falling apart and to add flavour. If the coating is sticky do not use a basket as the food will stick to it.
- Do not overflow or overfill the basket. This may drop the oil temperature too much and not allow room for movement of food.
- Drain food on absorbent paper before eating.
- Generally foods that are deep-fried need to be served and consumed immediately.

Safety

Deep frying can cause problems in the kitchen if not managed properly. Never leave the fryer unattended and ensure cords and so on are placed away from harm. If the frypan catches alight, cover the pan with the lid or fire blanket – **do not** use water.

Food examples

Foods best suited to deep-frying include:

- crumbed chicken or fish
- spring rolls
- battered fruits or vegetables
- doughnuts
- croquettes.

Did you know ...

Beer is a popular addition to batters because the protein in beer produces the browning effect and also provides a light, crisp, dry batter after cooking.

Recipes for deep-frying

Spring Rolls
(Makes 20 small)

50 g cellophane or vermicelli noodles
5 large spring roll wrappers, cut into quarters
4 spring onions, finely chopped
1 carrot, peeled and grated
150 g beef or pork mince
1 tsp sugar
1 c bean sprouts, trimmed
¼ c water chestnuts, finely chopped
1 tsp flour
oil for deep frying
¼ c sweet chilli sauce, for dipping

1. Soak the noodles in hot water until soft.
2. Prepare vegetables.
3. Drain noodles and chop into shorter lengths.
4. Mix vegetables, noodles and mince together.
5. Mix the flour with a little water to make a paste.
6. Place about 1 T of mixture along bottom edge of wrapper, leaving enough space at sides to fold wrapper over. Fold the sides in towards each other and firmly roll up. Seal edges with flour mixture.
7. Heat oil in deep fryer until hot. Place spring rolls in oil being careful not to overcrowd. Cook until golden brown
8. Drain on absorbent paper.
9. Serve with sweet chilli sauce.

Fritters
(Serves 2)

Batter:

1 c plain flour	30 g butter, melted
200 ml milk	2 egg whites
2 egg yolks	

1 apple	2 tsp sugar
1 banana	Deep frying oil
½ tsp cinnamon	

1. Prepare the batter. Sift flour into bowl and make a well in the centre.
2. Ensure the eggs are separated.
3. Add the egg yolks and half the milk to the well and using a wooden spoon gradually work in the flour.
4. Gradually add the remaining milk, beating well between each addition to make a thick batter.
5. Stir in the melted butter.
6. Stiffly beat the egg whites in a clean bowl, then fold into the batter mixture.
7. Core apple and slice into ¼ cm thick rings. Cut banana into 4.
8. Heat oil in deep fryer.
9. Coat apple and banana in batter, and deep fry when oil is hot. They may need turning for even browning. Remove from oil when golden brown and drain on absorbent paper.
10. Toss in combined sugar and cinnamon and serve hot.

Activity 8.6: Oil absorption in food

Research the amount of absorption of oil in relation to the thickness and size of a piece of food, for example potato. Peel and cut potatoes into varying sizes, from French fries to steak chip sizes. Using controlled variables such as the amount of oil, temperature, frying time and weight of potato, fry the foods and record the amount of oil left after frying. Note: This will have to be done after the oil has cooled down. What observations can you make from this activity? What implications does it have diet-wise? Discuss with your teacher and write notes here.

D. Grilling/Barbecuing

Description

This is a method of cookery that is fast and dry using heat radiated from a close source such as an element, open fire or hot coals. It is suited to small pieces of tender food requiring a short cooking time such as vegetables, fish and some meats. The fat released from the food is usually drained into a collecting tray situated underneath the food.

It is also a process used to brown or caramelise the surface of some foods such as mornays or crème brûlées.

Effects on properties of food

Foods are cooked quickly so that food is not exposed to heat for too long and more nutrients are retained. Food retains most of its colour and becomes drier as liquid is lost. Browning occurs quite quickly, sometimes turning black if cooked for too long. Aromas are generally pleasing. Food texture firms externally but remains internally moist if quickly cooked.

How to grill

This method is as per dry-frying except the fat produced should not have to be drained off. Use tongs to turn foods being careful not to damage them.

Equipment

The heat source for grilling is from either above (such as in domestic stoves) or below (as in barbeques). Other equipment may include salamanders (commercial kitchens) and electrical appliances such as vertical grills.

Pointers (addressing quality control):

- Be aware of fire concerns/hazards when grilling or barbecuing with open flames.
- Preheat the grill before using.
- If using open coals be aware that smoke will flavour the foods.
- Follow manufacturers instructions when using electrical or gas equipment; for example, sometimes doors have to remain open while grilling.
- Remove excess fats from foods where possible; snip the fat on chops and steak to prevent curling.
- Brush or spray some foods with a little oil to prevent sticking.
- Turn food carefully with tongs to prevent piercing and losing juices.
- Thicker foods take longer so adjust the temperature accordingly to prevent burning the outside of the food.
- Be careful not to overcook meats as this will dry them out and make them tough.
- Season after cooking as salt will draw out moisture and increase the dryness of the food if seasoned prior to cooking.
- Most grilled foods must be consumed immediately.

Safety

As outlined above, there are fire concerns and safety hazards with open flames. A type of fire extinguisher is advisable even when outdoors. Do not use water when hot fat is present. Be careful of burns from handling hot equipment and standing too close to an open grill. Use oven mitts and long tongs to minimise burns.

Food examples

Foods best suited to grilling or barbequeing include:

- cuts of beef such as scotch fillet, t-bone, fillet, porterhouse
- chops – lamb, pork
- chicken pieces if not too large, and of even sizes
- fish
- sausages
- bacon
- vegetables
- browning mornays, lasagnes or cheese topped dishes
- caramelising brûlées or other sugar topped dishes.

Did you know ...

A recent survey of UK shoppers has found that Australia's most recognisable food and method of cookery is the barbecue? It symbolises Australia's relaxed lifestyle, outdoor living, sunshine and general healthiness and well-being.

Recipes for grilling/barbecuing

Pork and Apple Kebabs

750 g diced pork
3 red apples, chopped
6 new potatoes
1 carrot, julienne
½ zucchini, julienne

Marinade:
15 g butter
1 Tbsp golden syrup
½ tsp grated lemon rind
¼ c brown sugar
¼ c apple juice
2 Tbsp lemon juice

1. Combine pork and marinade in large bowl, cover, refrigerate 2 hours.
2. Drain pork, reserve marinade.
3. Thread pork and apple alternately onto bamboo skewers.
4. Grill for about 6 minutes on each side. Brush occasionally with reserved marinade.
5. Serve glazed with any remaining marinade accompanied with steamed carrot and zucchini and baked potatoes.

Activity 8.7: Grilling

Find out what a hibachi is and which country uses it; what were its origins; and what types of food are suited to this piece of equipment.

WET COOKING METHODS

The wet methods of cooking employ moisture in their processes and can be used in a wide variety of foods. The methods typically use a liquid cooking medium, usually stock, water, milk or fruit juice. The liquid cooking medium transfers the heat to the food and cooks the food by convection.

A. Boiling and simmering

Description

Boiling is the method whereby food is cooked in water at a temperature of 100°C. At this temperature, large bubbles rise from the bottom of the vessel (the closest point to the heat source) and burst at the surface. Foods that are boiled do not dry out and do not brown. Cooking is generally evenly done as the whole part of the food comes into contact with the heated liquid.

Simmering is a slower and more gentle process than boiling whereby the temperature of the liquid is around 85°C and hence only small bubbles slowly break the surface of the liquid. This process is suited to soups and some dishes where the food needs to remain intact and be gently cooked.

Blanching is the process whereby foods such as vegetables are plunged into boiling water for a short period of time and then are immersed in cold water to refresh them. Foods that are blanched are often used in salads and retain their crispness while improving their texture.

Effects on properties of food

Food becomes soft and moist, absorbing the liquid in which it is cooked. Simmering will keep delicate foods intact. Water soluble vitamins are lost if liquids are drained away. Colour can fade over a longer cooking time. Foods can be served very hot.

How to boil

1. Fill saucepan with water to two-thirds full. It should be large enough to accommodate the food easily. Place onto heat.
2. While water is heating, prepare foods for boiling. See note about eggs.
3. Place foods into boiling water and bring back to the boil. It can be stirred; however, rice is best left alone to prevent damage to the grains.
4. Boil until food is cooked or tender.
5. Drain into a colander and rinse if required.

Note: Eggs should be placed into cold water and brought to the boil, not added to boiling water. This will help prevent the shell from bursting.

Equipment

For this basic method of cooking, saucepans, pots and stockpots are usually employed.

Pointers (addressing quality control):

- Ensure that the cooking vessel is large enough to prevent boil-overs. Remember liquid expands upon boiling and then adding foods to the liquid increases the whole volume in the saucepan.
- Boiling can be a rough process so is not suited to delicate foods.
- A rolling boil refers to the continuous movement of breaking bubbles on the surface of a liquid.

Safety

Try not to lift stockpots full of hot liquid. Open lids away from yourself to minimise steam burns and scalds. Boiling toffee mixtures present heat issues so use oven mitts to help prevent burns.

Food examples

Foods best suited to boiling and simmering include:

- vegetables
- rice and pasta
- cereal grains, pulses
- corned meats
- eggs
- soups and stocks
- shellfish
- jams, toffees
- milk.

Did you know ...

It is better not to rinse pasta after cooking it as rinsing removes the starch from the outside of the pasta making it slippery, and therefore sauces cannot 'cling' to it.

Recipes for boiling and simmering

Bechamel Sauce

(1 quantity)

600 ml milk
1 small onion, peeled
1 bay leaf
6 peppercorns
60 g plain flour
salt
white pepper
grated nutmeg
60 g butter

1. Place onion, bay leaf, peppercorns and milk in saucepan to scalding point (skin forms on top and a rolling boil is reached). Set aside.
2. Melt butter in another saucepan, remove from heat and blend in flour. Return to heat and cook until it forms a smooth golden paste. Remove from heat.
3. Add milk gradually to paste, stirring carefully at each addition to prevent lumps forming. Once all milk is added return to a medium heat and bring to the boil continuously stirring.
4. Simmer sauce for 5 minutes, then season with salt, pepper and nutmeg.
5. Use as required.

Chicken Stock

(Makes 4 L)

1.5 kg chicken bones
2 onions, halved
2 carrots, halved
2 celery sticks, chopped
8 L water
8 black peppercorns
4 stalks flat-leaf parsley
2 bay leaves
1 Tbsp salt

1. Preheat oven to 200°C.
2. Place chicken bones, onions, carrot and celery in a large baking dish and cook for 1 hour or until well-browned.
3. Remove from oven and transfer to a stockpot.
4. Add remaining ingredients and bring to the boil, allowing to simmer rapidly, uncovered for 10 minutes.
5. Reduce heat to a slower, constant simmer for 1½ hours.
6. Skim surface to remove impurities, and strain through a fine sieve.
7. Allow to cool and remove any solidified fat from the surface.
8. Refrigerate for up to three days, or alternatively freeze in portions.

Egg Custard

1 c milk
1 c cream
1 vanilla bean, split
5 egg yolks
½ c caster sugar

1. Place milk, cream and vanilla bean in heavy based saucepan and bring to a simmering point.
2. Whisk egg yolks with the sugar until light and foamy.
3. Gradually pour in warm milk and cream whisking continually.
4. Return to clean pan and cook over moderate heat stirring continually until mixture thickens and coats the back of a wooden spoon.
5. Strain into a cold basin. Serve warm or cold.

Activity 8.8: Boiling food

Test out the suitability of boiling a variety of foods. What common characteristics do the successful and unsuccessful food types have? What recommendations would you make from this? Try foods such as pasta, bread, beans and flour. Use your imagination!

B. Poaching

Description

This cookery method involves a flavoured liquid being heated in a shallow pan to not quite boiling or simmering point (80°C) and food being placed in the liquid so that it is only half covered. In this method, the foods are very gently cooked and retain their shape well.

Effects on properties of food

Poaching is a gentler process than boiling/simmering and foods tend to retain their shape yet become soft. Water soluble vitamins are lost. Colour and flavor can be determined by the liquid medium.

How to poach

1. Prepare the poaching liquid and place into the saucepan/frypan to heat slowly.
2. Prepare foods by carefully trimming, peeling or cutting.
3. Once the poaching liquid is simmering slowly, place the food into the liquid.
4. A cartouche may be used to keep the food immersed in the liquid. This is a circle of greaseproof or baking paper cut to the size of the vessel and placed onto the surface of the liquid over the food during cooking.
5. Poach until the food is tender. A lid is not necessary; however, keep an eye on the liquid level so that it doesn't reduce too much and reach a boiling point.
6. Remove food from the liquid carefully using a slotted spoon and serve immediately.

Equipment

This process can be done in wide saucepans, frypans or even specialised equipment such as egg poachers. Egg rings are also used in cooking poached eggs to assist the egg white to remain close to the centred yolk. Slotted spoons aid in the removal of foods from the poaching liquid.

Pointers (addressing quality control):

- Make sure the water remains at the correct temperature. Boiling may distort the shape of the foods.
- Use correct cooking times to prevent the food becoming overcooked and rubbery.

How to poach an egg:

1. Use a little butter or spray oil to grease the inside of an egg ring.
2. Break egg into a saucer or small vessel.
3. Place about 1 cm of water in a frypan so egg will not be covered.
4. Bring water to boil then reduce heat to a low simmer.
5. Place ring into water and gently pour egg into ring.
6. Egg white should start to set straight away. Be careful not to let the water boil or simmer too fast or the white will become tough.
7. Gently spoon the water over the yolk until it is set enough to your liking.
8. Remove with a slotted spoon and serve immediately.

Safety

As with all methods using hot liquids and equipment be careful to prevent burns from occurring.

Food examples

Foods best suited to poaching include:

- fruits; for example, pears
- eggs
- fish
- chicken.

Recipes for poaching

Poached Pears
(Serves 2)

2 large pears
125 ml red wine
125 ml water
½ c sugar
1 cinnamon stick
1 strip lemon rind
1 tsp cornflour, optional

1. Place sugar, wine and water in large saucepan. Heat gently over low heat until sugar is dissolved.
2. Wash and peel pear very carefully so not to damage flesh. Leave stalk attached.
3. Bring syrup to the boil and add lemon rind and cinnamon.
4. Lower heat to a gentle simmer and poach pears until translucent, around 20-30 minutes.
5. Thicken syrup for serving if desired by dissolving cornflour in a little water and adding to syrup. Bring to the boil, stirring continuously until thick and clear.

Activity 8.9: Food colouring and flavouring

1. Use different coloured liquids to colour foods such as pears. Is there any problem eating a food that is not coloured to our expectations? What reactions do people have when faced with differently coloured foods? Test this out on class members. Use a recipe with a sweetly flavoured poaching liquid to poach pears and then colour it differently to a person's expectations and record their reactions.
2. What are some flavourings that you could add to the poaching liquid and what foods would they be suitable for? Think of both savoury and sweet options.

C. Steaming

Description

This is a healthy and generally quick method of cooking that utilises the steam created from boiling water to penetrate and cook foods. The food remains complete with excellent colour and flavour retention. This process does not brown foods, rather it enhances original colour making it a great option for vegetables.

Effects on properties of food

Steaming is considered a healthy form of food processing because the food is in minimal contact with water so water soluble vitamins are generally retained. Colour is excellent and food can be served very hot. Texture can be crisp (vegetables) through to soft (puddings).

How to steam

1. Prepare foods for steaming by trimming, cutting and so on.
2. Fill base of steamer or saucepan to one-quarter full of water. Bring to the boil and then reduce to a simmer.
3. Place foods into the top steamer section or basket and cover with the lid. Place on top of the simmering water and cook until tender.

Equipment

Steaming requires the use of a double saucepan and lid with a steamer insert or, alternatively, a separately purchased steamer basket that fits snugly onto an existing saucepan. Electrical steamers can also be purchased. Some manufacturers offer a bench-top steamer component as an optional part of your cooking package, like a grill or a deep fryer!

Pointers (addressing quality control):

- Do not overcrowd the steaming basket. Cook in batches if necessary.
- Ensure there is no leakage of steam to prevent water loss in the saucepan.
- Be careful not to overcook and create limp foods.

Safety

Steam can produce nasty burns so be careful to remove lids away from your body.

Food examples

Foods best suited to steaming include:

- fish
- fruits
- vegetables
- puddings
- Asian foods, dim sum, pork buns.

Recipes for steaming

Golden Syrup Steamed Pudding
(Serves 4-6)

125 g butter
125 g sugar
1 egg, beaten
¼ tsp vanilla essence
1½ c SR flour
125 ml milk
1½ Tbsp golden syrup

1. Fill electric frypan half full of water. Bring to the boil.
2. In a bowl, cream together the butter and sugar. Add egg and vanilla, mix well.
3. Alternatively stir in milk and flour mixing thoroughly.
4. Grease a 1.5 L pudding basin and line with alfoil. Grease alfoil.
5. Place golden syrup in bottom of basin.
6. Place mixture into basin two thirds full and cover with foil overlapping mould, secured in place with string or elastic bands.
7. Place in frypan and steam with water simmering rapidly. This should take 1 hour. Test with skewer.
8. To serve turn onto warmed plate.

Activity 8.10: Steaming and flavouring

Try to be creative and see if flavouring the steaming water produces different effects on the steaming foods. Maybe use liquorice immersed in the steaming liquid to see if it will flavour a bland vegetable such as peeled, cubed zucchini.

D. Stewing

This method of cookery is suitable for tougher cuts of meat and foods as it utilises a long, slow cooking time with the presence of a liquid. This combination is excellent in softening food. The food is cut into smaller pieces and placed in a small quantity of liquid, often flavoured, in a saucepan and covered. It is placed on top of the stove and allowed to simmer gently until the food is tender. The food is usually served in its entirety including the sauce created from the liquid and any other foods added throughout the cooking process.

Effects on properties of food

The long cooking process involved in stewing is not beneficial nutritionally, with nutrients being exposed to heat for longer periods thus destroying them more. Tough pieces of food soften, absorbing liquids used and becoming more tender over the longer cooking process. Food tends to lose some colour, with **dextrinisation** and **caramelisation** creating a browning effect.

How to stew

Because of the varying nature of the foods/dishes produced by stewing, methods may vary. Consult your individual recipe for the process required. A general method is as follows:

1. Trim meat and cut into cubes.
2. Toss meat in seasoned flour.
3. Heat oil in heavy based frypan and add meat in small quantities, browning on all sides.
4. Add other ingredients such as flavourings and onions and liquids and cook slowly, gently simmering, covered, for approximately 1 hour depending on the cut of meat used.
5. Add prepared vegetables and cook covered until soft.
6. Serve with pasta or potatoes.

Equipment

Heavy-based saucepans with lids are suited to this process.

Pointers (addressing quality control):

- Keep food pieces roughly the same size where possible.
- Do not overcook – meat will become stringy and vegetables will be too soft.
- Ensure enough liquid is present for the cooking process.
- Make sure the cooking vessel has a heavy base to prevent food from catching on the bottom.

Safety

As per poaching and other wet methods. Be careful of heavy saucepans when lifting them off the stove.

Food examples

Foods best suited to stewing include:

- meat – lamb ragout
- chicken – chicken fricassee
- fruit – apple, rhubarb, apricots.

Recipes for stewing

Beef Stroganoff
(Serves 4)

750 g rump steak
¼ c plain flour
1 tsp paprika
60 g butter
1 large onion
1 clove garlic
250 g fresh button mushrooms
1 Tbsp lemon juice
2 Tbsp dry red wine
2 Tbsp tomato paste
1½ c sour cream
2 Tbsp chopped fresh chives

1. Trim and cut steak into thin slices. Toss in combined flour and paprika.
2. Melt butter in heavy based saucepan. Gently fry finely chopped onion and crushed garlic until golden brown.
3. Increase heat and in batches cook steak in saucepan stirring constantly until it is all browned.
4. Add mushrooms, juice and wine and stir until ingredients are combined.
5. Reduce heat, cover and simmer over low heat for about 5-10 minutes or until steak is tender.
6. Stir in tomato paste and sour cream. Stir constantly over a medium heat until mixture is heated through.
7. Serve with boiled pasta or rice, and steamed greens. Sprinkle with chives before serving.

Activity 8.11: Stewing

Devise a set of recipes for a slow-cooker for the busy household. Include a variety such as vegetarian. What foods would be unsuitable for this long process and why?

E. Casseroling

Description

Casseroling is similar to stewing except the cooking process is completed in the oven using an ovenproof dish. The meat is usually browned on top of the stove in a little oil to seal in the juices and provide a little colour and flavour. The meat may also be rolled in a little flour before browning which provides a thickening agent to the cooking casserole.

Effects on properties of food

Casseroling has the safe effect of foods as stewing.

How to casserole

Use the same process as per stewing, however at Step 4, complete the cookery process in the oven.

Equipment

The main requirement for casseroling is an ovenproof vessel with a lid. This can be ceramic, china, glass or even metal dish such as enamel. Cast iron can be used both in the stove and on the top.

Pointers (addressing quality control):

- Foods should be of even size.
- Meats are usually browned before being put into the casserole dish.
- The liquid forms part of the dish and requires thickening in the process.
- Some ingredients of the dish may have to be added during the cooking process to prevent their overcooking.

Safety

Be careful when removing hot heavy dishes from the oven. The contents may also spill if very full so ensure the vessel is not too full before it commences cooking.

Food examples

- Lancashire hot pot
- chicken creole
- apricot chicken.

Recipes for casseroling

Chicken Pilaf
(Serves 2)

125 g chicken, diced
1 tsp butter
½ onion, diced
½ capsicum, thinly sliced
½ celery stick, sliced
⅓ c rice
¾ c water
1 tsp chicken stock
salt
pepper
¼ tsp basil
4 parsley sprigs, chopped

1. Preheat oven to 180°C.
2. Melt butter in saucepan and sauté onion until a golden colour.
3. Add diced chicken, capsicum and celery and cook for 2 minutes.
4. Mix in rice and stir until it is coated with butter.
5. Add stock, water and seasoning, bring to the boil.
6. Place in a lightly greased casserole dish. Cover. Cook in oven until rice absorbs liquid, around 25-30 minutes.
7. Serve garnished with chopped parsley.

Activity 8.12: Casseroling

Create a set of casserole recipes for a person with a busy lifestyle. Include some vegetarian dishes as well. Recipes should serve two.

F. Braising

Description

This process is a combination of stewing and baking and is suited to tougher meats that require a slow, moist cooking method.

Effects on properties of food

Foods have a minimal loss of nutrients if served with the cooking liquid. Some fat is used. When foods are seared first, they have a brown appearance that may fade a little. Flavour can be manipulated depending on the cooking liquid used. Foods are generally softer yet retain shape well.

How to braise

1. Prepare a mirepoix of diced vegetables, usually carrot, onion and celery.
2. Heat a heavy-based saucepan (or frypan) and sauté vegetables with a little oil. Herbs can be added at this stage if required. Remove from pan.
3. If red meat is used, add to pan in small batches and brown evenly to seal. Remove each batch until all is done.
4. Return vegetables to pan, then place (browned) meat on top.
5. Add a liquid, usually a stock flavoured with a little wine, and cover. Liquid should be enough to partially cover the food.
6. Cook gently until meat is tender and, when pierced, juices run clear.
7. Serve. The vegetables are not used in the final dish, however a tasty stock could be created from the remains in the pan.

Equipment

Use a heavy-based saucepan or frypan with firm covering lid.

Pointers (addressing quality control):

- White meats such as poultry are sometimes not browned and white wine is used to flavour the stock.

- Red meats require browning and use a red wine to flavour the stock.
- Keep an eye on the liquid level to prevent foods from drying out and sticking to the pan.
- Keep even proportions of the mirepoix vegetables to prevent one from overpowering the others (be careful with strong-flavoured vegetables like onions and parsnip). The mirepoix adds flavour to the cooking meats.

Safety

With braising there are burn issues with steam and hot vessels. Be careful to minimise these dangers. Also, the lifting of heavy pots can cause injury if not managed correctly.

Food examples

Foods best suited to braising include:

- braised steak
- braised vegetables such as celery, turnips, witloof
- Risotto.

Recipes for braising

Braised Steak and Mushrooms
(Serves 6, Cooking Time: 2-3 hours)

500 g whole round or topside steak
2 carrots
2 brown onions
3 celery stalks
½ turnip
40 g butter
bouquet garni
1 L beef stock
2 Tbsp red wine
300 g mushrooms, sliced
30 g butter, extra
1 Tbsp lemon juice

1. Peel and mirepoix carrot, onion, celery and turnip.
2. Heat butter in large heavy based saucepan and add vegetables. Fry lightly until browned.
3. Add meat and cook, turning once until browned.
4. Add stock to cover vegetables, cover and simmer gently for about 1 hour. Add wine and simmer for another hour adding more stock if required.
5. Remove meat and keep warm. Strain juices and return to saucepan. Cook, reducing until a thickened gravy is formed.
6. Heat butter and lemon juice in small frypan. Add mushrooms and sauté until tender.
7. Slice steak and serve with gravy and mushrooms with separately cooked vegetables.

Braised Chicken Drumsticks
(Serves 2-3)

2 cloves garlic, chopped
1 onion, diced
2 large red chillies, seeded and finely diced
¼ tsp ground cumin
¼ tsp ground coriander
¼ tsp ground turmeric
¼ tsp five spice powder
1 tsp shrimp paste or 1 T fish sauce
4-6 chicken drumsticks
2 Tbsp oil
125 ml chicken stock
125 ml coconut milk
1 Tbsp fresh coriander, shredded

1. Grind or process garlic, onion, chilli, spices and shrimp paste to form a thick paste.
2. Remove the skin from the chicken.
3. Heat oil in a heavy based frypan, add the chicken and cook until browned all over. Remove and keep warm.
4. Add the paste to the frypan and cook over a low heat 1 minute.
5. Return the chicken to the frypan with the stock and coconut milk. Cover and simmer for 30 minutes or until chicken is cooked through, stirring occasionally.
6. Serve garnished with coriander.

Activity 8.13: Braising

You have a surplus of chicken legs. Describe how you could use these in a braising method and what you would serve them with. What factors do you have to consider when planning your dish?

MICROWAVE COOKERY

Description

This is a quick method of cooking suitable for a large variety of foods. To microwave foods you require a specialised oven that uses a magnetron to convert electrical energy into microwaves or electromagnetic energy. These waves are not visible and travel into the oven from the magnetron whereby they are dispersed by a fan to scatter them through the oven as evenly as possible. The waves pass into the food at a depth of 4 cm and act upon water molecules vibrating them and causing friction. Friction energy is then transferred into heat energy which is conducted through the food and hence the food is heated and cooked.

Effects on properties of food

Depending on the amount of cooking time, foods can be firm or soft. If water is used, water soluble vitamins may be lost. Colour retention is very good, however foods cannot brown in the cooking process.

Did you know ...

Microwaving is either a wet method if liquid is used in the food being microwaved or a dry method if no liquid is added to the foods being cooked.

How to microwave

It is best to follow set microwave recipes when cooking in the microwave oven. Your oven will come with a detailed set of instructions outlining simple defrost techniques and hints for simple food preparation. Some also come with a separate recipe book and it is useful to try this out to become familiar with what the microwave oven can do.

Microwave tips

- Melt cooking chocolate on 50% power for 1 minute. Stir then repeat in 30 second timings until almost melted. Chocolate will still melt upon standing.
- Crystallised honey can be recovered by cooking in the microwave on 50% power. The time depends upon quantity. Transfer honey to a suitable container that won't melt in the microwave oven.

Equipment

Basic equipment requirements include the microwave oven and containers suited for the microwave such as some plastics, glass and ceramics. Some materials reflect the waves rather than absorb them so are not suited to the microwave, such as foil and metal.

Pointers (addressing quality control):

- Become familiar with the manufacturer's instructions before using the microwave.
- Ensure maintenance and regular checks are undertaken for safety.
- Foods with higher water content cook faster than those with lower water content.
- Some foods cannot be cooked in the microwave, such as eggs in their shell and shellfish. Those foods that require crisping or browning are also not suited unless the oven has a convection function.
- Make sure food is allowed to stand before removing it from the oven to continue the cooking process.
- Potatoes will need to be pierced or pricked to allow steam to escape whilst cooking.
- Frozen foods need to be thawed completely before cooking so always defrost on the recommended setting. Incomplete thawing before cooking can increase the issue of food remaining in the 'danger zone' for food contamination to occur.
- Use paper towel instead of plastic film to cover foods. This minimises the chances of contaminants from the breakdown of the plastic film coming into contact with the food.
- Place foods evenly around the turntable to enable even cooking.

Purchasing a microwave oven

These appliances are becoming more and more affordable. Points to look for when purchasing are:

- oven capacity
- wattage or power
- reliable manufacturer
- warranty period
- power levels
- defrost function
- turntable present
- inverter capability
- convection ability if you require a browning function.

Did you know ...

You can disinfect a kitchen sponge by placing it in the microwave oven for 60-120 seconds. The smell is also improved.

Safety

Problems can occur with burns and steam burns. Be careful when removing hot objects from the oven; use oven mitts and remove coverings away from you.

Food examples

- vegetables
- precooked, packaged meals
- pasta and rice dishes
- defrosting foods
- desserts such as puddings, custards
- fish
- chicken
- popcorn
- hot chocolate.

Recipes for the microwave

Vegetable Risotto
(Serves 4)

1 onion, diced
2 celery sticks, sliced
30 g butter
1 c long grain rice
125 ml dry white wine
500 ml chicken stock
½ red capsicum, diced
½ green capsicum, diced
4 spring onions, sliced
1 Tbsp chopped fresh parsley
¼ tsp turmeric

1. Prepare vegetables. Grease a large shallow dish.
2. Place onion, celery and butter in dish and cook covered on HIGH for 3 minutes.
3. Add rice, wine and stock and cook for a further 15 minutes on 70% power or until rice is tender. Stir occasionally during cooking.
4. Add capsicum, spring onions, parsley and turmeric. Cook on HIGH 2 minutes or until heated through. Allow to stand 2 minutes.
5. Serve hot.

Activity 8.14: Microwaves

There has been some controversy over the use of microwaves in cooking. Find out what the issues are and, using a variety of publications, make up your own recommendations and thoughts on microwave cookery. Summarise your findings.

REVIEW

Review Activities

1. Fill out the following table as a review.

Heat transfer	Definition	Relevant cooking methods	Cooking equipment
Convection	Heat passes through another medium.	P ____________ B ____________ S ____________ B ____________ B ____________ R ____________	
Conduction	Heat is transferred to the food by direct contact with the cooking vessel.	S ___- F _____ S ______- F ______ Sautéing	
Radiation	Heat transfers directly onto the food being cooked.	G ____________ B ____________ M ____________	

2. In the chart on the following page, the boxes are randomly placed.
   - Get your teacher to photocopy the chart for you.
   - Cut each square out and place them in the correct charting format. You will be given three minutes to do this.
   - Then go and look at others around you and you will then have another two minutes to place them in the correct sequences.

- Share with others how you came up with your sequencing.
- Glue it all together when you are sure of your answers.

Court Bouillon, stock syrup, cartouche	Boiling water, steam	Hot liquids, splashing	Large pot, stockpot slotted spoons, ladles, colander	Braising pan, braisiere, sauté pan	Eggs, fish, fruit, poultry
Cooking larger pieces of food where the liquid only half covers the food	Food is immersed in a liquid and cooked at 100°C.	Braising	Magnetron	**Related cooking terminology**	Slow, gentle moist heat method of cooking in which the food is covered by a liquid
Definition	Vegetables, soups, puddings	Meat, poultry, fish, seafood, vegetables, fruit	**Occ. Health and Safety related issues**	Corned meats, vegetables, eggs, pulses, shellfish	Microwave, glass and ceramic cookware
Vegetables, seafood, poultry, puddings	Deep pan, fish kettle, slotted spoons	**Cooking method**	Cooking of food in a liquid at just below boiling point	Radiation, steam, spillages	Boiling water. steam, heavy pots
Splashing, spills	Poaching	Boiling	Microwaving	Food is cooked in the steam produced by a boiling liquid	Radiation waves cause the food to first heat then cook.
Chicken, vegetables, tougher cuts of meat	**Suitable foods**	Deglaze, lard, basting	Stove, oven, bratt pans, pots, ovenproof dishes, wooden spoons	**Utensils and Equipment**	Steaming
Parboiled, stock, simmer, blanch, refresh, al dente	Compote	Pressure cookers, steamers	Hot splashes, heavy pans, steam	Stewing	Atmospheric steaming, high-pressure steaming

3. In your own words, define the following terms: convection, conduction, radiation, moist heat cookery, dry heat cookery.

4. Why are deep and shallow-frying classified as a dry method of cooking, when food is cooked in liquid fat?

5. Give an example of food cooked in the microwave using moist heat. Give an example of food cooked in the microwave using dry heat.

6. Describe the type of heat transference used for the following:
   - roast beef and turned vegetables
   - pumpkin soup
   - pan-fried chicken fillet
   - poached pears.

7. Outline five safety points to be aware of when using wet methods of cookery.

8. Look at a recipe for Macaroni Cheese with Spring Vegetables. Identify the cooking methods it uses and then suggest alternatives for each process if you can. What consequences will there be for utilising these other cookery methods on the final product?

9. Investigate the cookery method of pressure cooking. Present the information in the categories outlined below.
   - Description
   - Equipment
   - Pointers
   - Safety
   - Food examples
   - Basic set of recipes

EXTENSION

Extension Activities

1. Thinking about the foods that you enjoy, list the cooking methods they utilise. What makes these foods enjoyable to you? Why do you like them so much? Think about the healthiness or nutrition of these methods. List the positive consequences and negative consequences of consuming these types of foods (and hence cooking methods). What recommendations should you make for yourself and your daily diet?

2. Create an assessment tool and practise your proficiency at cookery methods. For each method and recipe tested, create a checklist to ensure you have met the criteria for a satisfactory result. Research what requirements or characteristics each recipe and method should display. Test your checklists out on other classmates and evaluate their effectiveness.

Useful Websites

www.taste.com.au: A range of recipes with an easy to use format plus techniques and information on foods.

www.allrecipes.com.au: A home cook recipe site for Australians and New Zealanders.

NOTE: There are a lot of Internet recipe sites – experience will eventually let you know which ones are good and which ones to avoid!

CHAPTER 9
Product Development and Evaluation Techniques

Key Concepts

- **Developing food products**
  - **Steps in developing a food product**
- **New and emerging products, systems and services**
  - **Innovation and the domestic kitchen**
  - **Innovation and commercial enterprises**
- **Testing and analysing food products and processing systems**
  - **Evaluation techniques**
  - **Tools for evaluating**

DEVELOPING A FOOD PRODUCT

Whether in a domestic situation or a commercial enterprise, the development of a new food arises from the identification of a need by consumers. Think about some of the foods that have emerged over the years in our supermarkets. A lot of them pertain to providing foods that are partially or fully prepared for the consumer to purchase, based upon the fact that people have busy lifestyles and are spending less time in the kitchen preparing foods. This type of food product has been successful with the evidence in people's shopping trolleys every week.

Examples of these include:

- frozen dinners
- pre-made pastry
- baby food
- quick serve noodles
- muesli bars
- snack items such as packaged cheese and dips
- pre-made fresh raw pasta.

A **need** was found by manufacturers that consumers **want** to spend less time in the kitchen. Therefore the products have been developed and now form part of the weekly shopping list for many consumers.

Another need was the increasing awareness of those suffering from celiac disease (or gluten intolerance). More and more gluten-free food products are arriving on supermarket shelves in a response to this growing market.

Manufacturers spend a lot of time researching consumer needs and wants, collecting information and analysing data. From this they can determine what is required and develop products based upon that data. By putting a lot of effort into research and development, manufacturers can save time and resources in making sure what they are developing will have an end market. Why would someone put time, effort and money into developing something that wasn't required? That would be a waste of resources and an economic disaster for those involved.

STEPS IN DEVELOPING A FOOD PRODUCT

This process bases itself upon the **technology process** as outlined in **Chapter 4**. Follow the steps and apply the principles of the technology process.

STEP 1 **Generate an idea**. This can be something you have identified as a need that is not present in the current market. Or it may be something you would like to see trialled.

STEP 2 **Is it a viable idea?** Will your product be economically and technically worthwhile? Ask for people's initial comments on your proposed idea and get their feedback. You may have to modify your idea to cater for their responses.

STEP 3 **Complete experimental work on the idea.** This has to be very controlled and accurate, especially when working with recipes. Write down in detail what it is you are trying to do with what. Treat it like an experiment and take notes! For example, follow this procedure:

i. Write an aim.
ii. Outline the procedure.
iii. Complete an accurate list of materials and quantities.
iv. Complete the procedure.
v. Record results.
vi. Make recommendations for further experimental work.

STEP 4 **Taste testing.** Once you have finally completed your experimental work, and have come up with an acceptable product, trial it on other people. A taste panel usually compares products to each other. These can be either products you have made yourself and are comparing slight differences of ingredients or cooking processes, or they can be already available commercial products of which you are trying a variation. A taste testing needs to be controlled with limited interaction between the taste testers and no identification of which food is what. These may sway the tester's responses and therefore your results will become invalid. Provide a glass of water to cleanse the palette between tastings and lessen confusion over the flavours.

A sample form is provided below for tasting results.

Sample No.	Appearance	Colour	Flavour	Texture	Comments
A					
B					
C					

A rating can be provided for each to give you a measurable evaluation. For example, 1= Dislike intensely, 2 = Dislike, 3 = Neutral, 4 = Like, 5 = Like very much.

You can also ask questions from the tasters as to what would they improve; would they purchase the product; how much would they pay for the product; and so on.

STEP ❺ **Draw conclusions.** At this stage of the product development, you need to take the results from the taste testing and determine whether or not you should go further with the product. Obviously if there were favourable responses from the taste testing it would appear that the product could go ahead. If there were some negatives, modifications would have to be made, the product would need to be refined and the above procedures would have to be repeated.

STEP ❻ **Develop the marketing of the product.** This involves the packaging and the labels. Refer to Laws and Regulations in **Chapter 11**, for the legislation and information for Australian packaging requirements. At this stage you can also seek outside information as to the product's appropriateness and appeal. Marketing has huge implications on a product's success, as much as does the product's appeal to consumers. Remember to consider environmental implications when packaging your product.

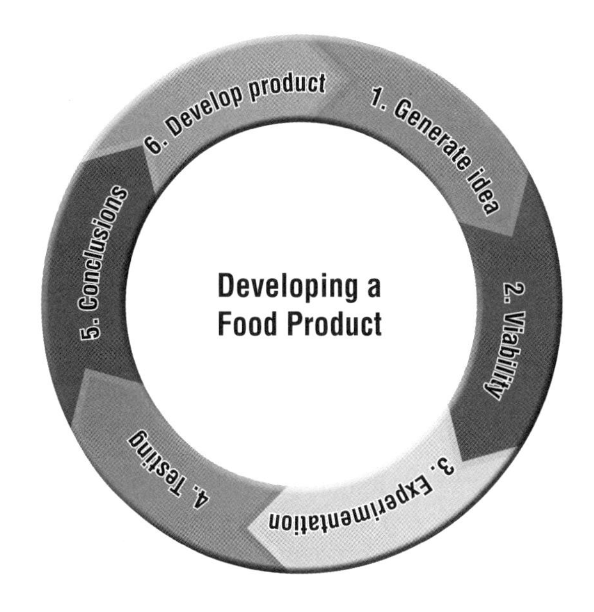

Activity 9.1: Design your own

Complete the design process above to come up with a new flavoured biscuit, bread or similar. Try it out in class and evaluate each other's.

NEW AND EMERGING PRODUCTS, SYSTEMS AND SERVICES

The Food Industry covers a very broad spectrum and when looking at innovation and the food industry, it could cover aspects from the growing stages of food to food processing to marketing and supply of food products, and then the hospitality industry becomes another whole area to investigate. To simplify this section the focus will be on equipment used and ingredients available to the domestic market.

Did you know ...

Some of the innovations occurring in the food industry are in the areas of:

- bioscience
- enzymology
- food microbiology
- food polymer and structural science
- food safety management and hazard prevention
- key supply chains
- process technologies
- shelf life extension strategies.

INNOVATION AND THE DOMESTIC KITCHEN

To fully appreciate the evolvement of technology over the years, you simply need to visit a museum and see what it was like for the early settlers to complete the process of cooking. You could even go back one generation when microwaves were only fledging and induction cook tops unheard of. In the kitchen, technology often involves small electrical appliances that simplify tasks for you. To capture market interest they are often quite a novelty, for example a popcorn maker. Some examples of these include:

- vacuum sealers
- pie and pizza makers
- low-fat deep-fryer
- full automated coffee machines
- steam ovens
- milkshake makers
- rice cookers
- food processors such as Thermomix
- ice cream maker.

Utensils can also be subjected to technology with many new gadgets being made for the domestic chef. A few recent examples of these are:

- the powerlance peeler that peels all fruit and vegetables, slices cheese and shreds cabbage
- the apple corer that peels, cores and slices in the one motion
- battery-operated pepper grinders
- electronic scales
- hand-held stick blenders
- butane torches.

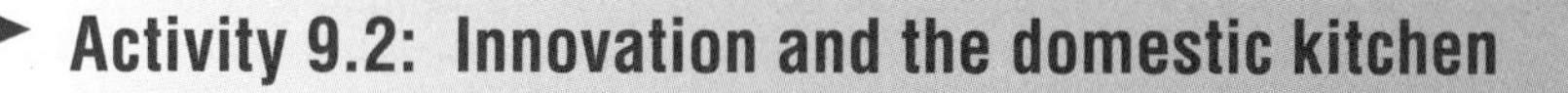

Activity 9.2: Innovation and the domestic kitchen

1. Think of your kitchen at home and describe some of the more innovative utensils that you have.

2. Within small groups brainstorm ideas that could be turned into innovative gadgets within the kitchen. Think of tasks that are done manually and try to think of some other way the process can be done. Outline your ideas and draw sketches and label the final invention.

3. Compare the production of making a sponge cake, initially with modern kitchen equipment and then alternatively with simple utensils. You may have to use a modern oven to cook the hand-made product. What aspects could you compare between the two products? Make up a comparison chart and finally explain the pros and cons for both methods.

Materials used in the kitchen have also been an area of continual development. Gone are the aluminium saucepans and in their place are high technology materials as well as a few of the traditional favourites.

Activity 9.3: Kitchen equipment

1. Cite examples of equipment and utensils that utilise each of the materials below:
   - aluminised steel
   - bakelite
   - carbon steel
   - cast iron
   - copper
   - nylon
   - silicon
   - stainless steel
   - stoneware

2. Find out what concern existed with aluminium and teflon being used as materials in the food industry.

INNOVATION AND COMMERCIAL ENTERPRISES

Commercial innovation is geared towards meeting consumer needs as quickly, as accurately and as most economically viable as possible. Most businesses operate as a means to make profits to satisfy their stakeholders and ensure their ongoing future in their industry. This often forces businesses to be proactive in seeking new technology to keep abreast of developments and not fall behind their competitors. This encourages the evolvement of new ideas and new products for consumers to try out. After all, consumers are always willing to add variety to their everyday lives and, domestically, trying out new foods or new food experiences are ways to satisfy this desire.

Examples of commercial innovation

There is a saying that 'you're only human' which refers to the inability of someone to be absolutely perfect. You may find yourself that you perform differently on some days compared to others. A pasta carbonara that you made last week was beautiful, however the one that you attempted today didn't taste the same, didn't have the same texture and didn't look the same. That is because humans create variables in any given situation. They may not do the skill to the same degree, they may make errors or mistakes, or there may be outside influences that affect the quality of the finished product such as fire drills interrupting the preparation and cooking process.

Businesses get rid of these variables to a degree with the introduction of **automation** into their factories or premises. This reduces waste, products become standardised and in the long term it is economically better for the manufacturer.

Production line – This is the chain of consecutive production phases or tasks that have to be completed to produce a final product.

Automation – This is the technique of making a device, machine, process or procedure more fully automatic. It is the act of replacing human skills with machine skills, for example self serve checkout systems.

Did you know ...

The world's ultimate commercial coffee machine is here. Apart from being huge, it has a DVD player with three flat screen TVs and surround sound system all built into it! The coffee tasting may take you all day to get through...

Other examples of currently emerging food technology innovation are:

- developing value-added products such as organic acids
- creating biodegradable plastics from food waste
- high pressure processing
- ultrasonics and food processing
- pulsed electric field technology

- cool plasma microwave technology
- wholefoods/organics development
- information technology incorporated into the kitchen – benchtops or appliance doors with computer screen interface capabilities.

TESTING AND ANALYSING FOOD PRODUCTS AND PROCESSING SYSTEMS

One of the more interesting and informative activities you can do is to compare a product that is made in a factory with one that is made in a domestic kitchen. There are obviously benefits to both processes based on appearance, taste, availability, production, quality and so on.

Evaluation techniques

When assessing a product, you need to be quite thorough and explicit in your recording of information. When asked what a sushi roll tastes like, it is insufficient to say 'good'! To give quality feedback you need descriptive and accurate information about what is being asked of you. You also need to be thorough with no information is 'left in your head', so that someone else could fully interpret what you are trying to say without any assumptions being made.

Activity 9.4: Evaluation techniques

Come up with about three to five words that could be used instead of the following. (Hint: use a thesaurus)

- Good
- Tasty
- Nice
- Brown
- Tasteless

In the process of **food preparation** there are many aspects that can be considered when evaluating. You will need to determine the most important criteria that you need to focus on. The criteria will need to be predetermined before the evaluation takes place so that accurate observations can be made. To do this, a checklist is often a helpful tool for evaluations. A checklist can simply be a tick process that indicates something has been done or completed; or it can be a measuring tool giving a scale of achievement, for example: 'Not done', 'Satisfactory', 'Well done', 'Excellent'.

Such a measuring tool is subjective and is best used with a set of standards for each aspect evaluated.

Aspects to consider may include:

- equipment usage
- food handling practices
- planning
- preparation techniques
- recipe interpretation
- safety management
- self management skills
- teamwork skills (if applicable)
- time management
- work efficiency.

When assessing the use of **raw versus processed food products**, make sure you look at the freshness of the final product. By using semi-processed foods, often a compromise is made concerning the quality of the ingredients. If comparing raw versus processed, make sure the evaluation addresses the use of added chemicals in processed foods, plus the manufacturing processes that the ingredients go through (eg. heat application) which may affect their nutritional quality. You will need to look at:

- appearance
- aroma
- texture
- flavor
- nutritional quality
- ease of use
- cost
- quality of ingredients
- palatability
- packaging if applicable
- shelf life
- final useability.

Did you know ...

Manufacturers of food products love using salt to fix problems! Cornflakes would taste metallic without the use of salt, and crackers would be soft and stick to the roof of your mouth if salt wasn't added...

Tools for evaluating

As previously stated, evaluations can be easily made using observational checklists. Sometimes other methods may need to be used if the checklist is not suited to the process or final product. Other methods that may be utilised could include:

- product success, ie. sales or consumption based results
- self-evaluation
- observations
- anecdotal records
- project diary
- questionnaires
- surveys.

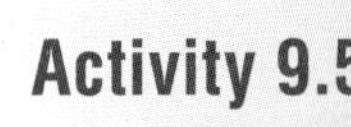

Activity 9.5: Commercial and home-made food

1. For each of the evaluation tools listed above, describe how you could utilise them in the production of a food system or product.
2. Outline the benefits of both commercially-made and home-made aspects of food production. Discuss with your class to get your responses.
3. Did you find out anything new? What are your preferences and why? Does your attitude change with different food products?
4. Devise a checklist to use in your next food practical lesson. Select at least five aspects of the food preparation list to evaluate. Decide whether you will use a tick-box system, or a measurement. Implement the checklist and explain how you could improve it to be used again.

EXTENSION

Extension Activities

1. Develop a range of after-dinner chocolates that you could manufacture in your home and sell to those wishing to entertain at home. Go through the steps in Developing a Food Product and trial it on your class. Decide whether or not you have come up with a viable product to sell.

2. Your group has decided that there is a need for a lunch package that is marketed at the preschool or primary aged group in your community. Go through the steps in Developing a Food Product and trial it out with a local primary school class. Decide whether or not you have come up with a viable product to sell to parents.

3. Develop a range of organic snacks that are aimed at the teenage market. Complete the steps in Developing a Food Product and trial it on the school population. Obtain feedback and determine whether or not you have come up with a viable product.

4. You decide there needs to be more awareness about the issue of the implication of food processing on the human diet. Develop a plan to make the general community more aware of the issue. Go through the steps in Developing a Food Product and display your results in the school library. Obtain feedback and make recommendations for this project.

5. You have decided that you wish to run a business from home, making food products to sell. Complete the steps in Developing A Food Product with a food that you could easily make, package and sell through a local food store. Write detailed notes on your venture and whether or not it would be a viable enterprise.

Useful Websites

www.can-news.com.au: Can and Aerosol News- look up "education" to find out about the canning process

www.sakata.com.au: How rice crackers are made. Technologies used.

www.foodaust.com.au: Food Australia, the official Journal of the Australian Institute of Food Science and Technology.

www.food-innovation.com.au: A website that promotes research and development in all aspects of food in Australia.

CHAPTER 10
Food in Society

Key Concepts

- **Factors that influence food choices**
- **Environmental issues that arise from food and lifestyle choices**
- **Influences on adolescent food choices**
- **Ethical influences on food choices, such as animal welfare, fair trade, resource use and country of origin**
- **Health issues that arise from food choices**

FOOD ISSUES

Food issues are concerns related to the supply, choice, cost, quality and sustainability of foodstuffs in today's world. People need food to survive and for their bodies to perform essential tasks. In striving to feed everyone in the community safely and nutritionally while also preserving the environment, concerns or issues arise related to the growing, processing, preparation and eating of food. Food issues affect the way people select food and the awareness of environmental concerns related to their food choice.

Factors that influence food choices

People choose what food to eat using many different criteria. A simple task of reaching for a snack at home may see you choose an apple for its health as well as economic aspects rather than the packet of crispy noodle snacks which look enticing but may have been imported (not supporting the Australian economy) and could also contain little nutritional value. However, if in the company of friends, that apple may be overlooked because everyone else in the group is eating something else and you may not want to appear too different. The following section outlines some of the variables faced when making food choices.

1. Cost

Some food choices are dependent entirely on cost. If there is a limited amount of money to spend on food, consumers will often go for the cheaper option. Being able to compare the prices of products required mental arithmetic by consumers until there was seen a need by the government to assist shoppers to make easier comparisons. The Australian Federal Government implemented the Trade Practices (Industry Codes Unit Pricing) Regulations 2009, enforcing all large supermarkets with a square space over 1000 m to display a unit price for all grocery items for which a selling price is displayed. The **unit price** – a price per unit of measurement (for example, per 100 grams) – must be displayed legibly, unambiguously, prominently and in close proximity to the selling price.

EXAMPLE

Cost: $10.74 for 250ml
Unit price: $4.30 for 100ml

Cost: $6.29 for 250ml
Unit price: $1.26 for 100ml

Activity 10.1: Unit price

Some products are exempt from unit pricing. Which ones are they? Research and list five examples.

With the unit price displayed, it is easy for the consumer to make a choice based on value for dollar.

The cost of the product can vary due to:

- **seasonality:** if not in season, storage costs may be passed onto the purchaser
- **demand:** when there is increased, companies can charge more for a product so they can maximise profits
- **competition:** if many types of the product are available, prices may be reduced so consumers are attracted to buy the lower cost product
- **importation:** if a product comes from overseas, it may face added transport costs as well as import duties and taxes
- **overhead costs:** are expenses added in to basic products including labour, packaging and advertising as well as business costs of the grower/producer.

2. Availability

The simple availability of the food can affect consumer's food choices. The availability of food can be dependent on the following variables.

- **Season of the year:** fruit and vegetables are in larger supply when grown in season and picked fresh. At other times of the year, they cannot be bought fresh unless imported; for example, stone fruit like apricots and plums are best in the summer months. They cannot be stored fresh for long periods of time, and need to be preserved if consumers want to eat them in winter.
- **Supply/demand:** if there is a sudden demand for a product and the producer/wholesaler cannot keep up with the supply, stocks will run out. An example of this is the mandarin juice that is sold at the Bunbury hockey carnival in June. It is such a popular product that stocks sell out almost immediately every day.
- **Natural disasters:** events such as droughts, cyclones, flooding and insect damage from plagues can affect crops, damaging them to an extent that there are reduced supplies for sale.
- **Industrial action:** strikes by workers involved in transporting produce can affect supplies, as can strikes by workers involved in manufacturing.
- **Man-made disasters:** train derailments or shipping accidents can affect the delivery of food items.

How we use the food that is available to us is also changing. With busier lifestyles, less time is available to spend at home making meals from 'scratch' – meaning preparing meals with little or no processed foods used in them.

How consumers cope with this aspect drives changes that are apparent in the marketplace.

i. Eating out

More people are eating out. Formal restaurants are still popular but even more so is 'casual dining' where either busy parents can take their children to eat out or a group of friends can get together and enjoy a range of foods tapas style. This has led to a change in market dynamics. The type of food served has also undergone a change with the concept of sharing food in the form of tasting plates now being increasingly popular. People can now have a drink and enjoy a simple meal using uncomplicated food – a concept more readily available in larger cities.

ii. Drive-through food outlets

These have been around for years, but the types of food available are increasing from the chicken and hamburger focus to others like doughnuts and tacos (especially in America). Some fast food chains are heavily promoting breakfast using the concept of a nourishing quick meal that won't make you late for work or study. There are even drive-through grocery stores being developed. The consumer need for instant food gratification means that the markets need to adjust their practices to provide it.

iii. Drink outlets

These are also increasing in popularity, especially in the areas of fruit juices and drive-through hot beverages. In busy shopping centres or retail areas, companies selling healthy fruit juices and wheatgrass shots are appealing to those wanting a drink packed with nutrients. Along busy highways, drive-through coffee outlets are positioned to allow the consumer to enjoy hot drinks like lattes, cappuccinos, macchiatos or even hot chocolates while they are driving to their destination.

iv. Cafés

These have undergone change with a lot of people going out and having a coffee – something that your grandparents would not have done in their younger years. There are many more coffee shops with more tables outside – usually on the pavement. Businesses concerned with cafés need to ensure they have the outside alfresco facilities to entice people who enjoy watching the world and, perhaps, being seen.

v. Breakfasts and brunches

These are becoming increasingly popular, and it is not uncommon for business meetings to be held over this mealtime instead of the more traditional 'long lunches'. Eateries that usually opened late morning have to adapt to this practice and open their doors earlier to catch this type of market.

vi. School canteens

These have also undergone change. In the past many operated for profit and sold foods that are now deemed unsuitable for children's health. Using the 'Traffic Light System', unhealthy foods like cool drinks have now been replaced with other beverages that are considered much healthier.

Even with so much fast food available, there was still the need to simplify meals that were cooked at home. This has led to supermarkets offering more processed foods and partially prepared meals that are wide-ranging in prices from simple home-brands to gourmet.

It is not just households wanting to simplify the process – even food outlets like cafés and some restaurants are looking to make things easier for themselves. Products like cakes and biscuits are not necessarily made on the premises but ordered from a food supplier who caters for food establishments in that area. That tasty slice of mocha chocolate gateau can be purchased from a café in Hillary's and an exact replica of it from Fremantle's coffee strip on the same day, both coming from the same supplier.

Consumers today also want healthy alternatives when choosing prepared foods. This has led to traditional chicken and chip stores or hamburger outlets increasing the salad options on the menu and providing skin-free or low fat items for consumers. Outlets selling healthy products have become very popular.

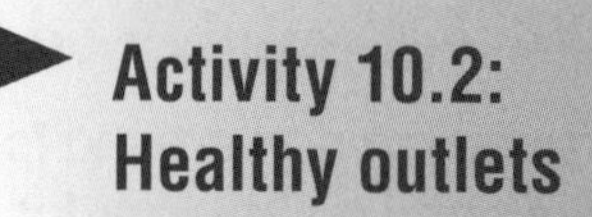

List examples of healthy takeaway food or beverage outlets in your area.

3. Family characteristics

The greatest single impact or influence on a person when growing up with regards to their food choices arises from their family. Up until young adolescence, the family has traditionally had the greatest influence on a young person's food habits. This can include simple things like what types of food are eaten, when meal times are, how food is cooked and where families eat. To get an idea of your family characteristics in regards to food selection, complete Activity 10.3.

Activity 10.3: Your food selection habits

1. **Daily foods:** describe the foods you have grown up with for the following:
   - breakfast
   - lunch
   - dinner
   - snacks
   - drinks
2. **Meal patterns:** think about your family and describe the typical meal time. Are you having your meal on the couch watching TV, are you all sitting around the table or is it like a 'grab and run' situation with people eating at different times? Has it changed over the years? Explain how this affects your food selection.
3. **Special equipment:** in recent years, the outdoor pizza oven has become popular, with many homes incorporating one in their back yard. This type of oven has been around for many years though and was especially popular in Perth's inner northern suburbs where there resides a large Italian population. Can you think of any food customs or cooking methods which you have in your family which you think are 'different' to other families or households?
4. **Food sources:** where does your food come from, that is, supermarket, home grown, orchards or wholesalers? Explain why your family obtains your food this way.
5. **Special events:** now describe some special occasions and any food customs that your family may have. Mention events like birthdays, Easter or Christmas if applicable, other religious occasions, anything!
6. **Takeaway foods:** these are becoming increasingly popular with families as the range and convenience of them grows. Think about how frequently you have takeaway as a meal or snack and describe what takeaway you buy and how often you eat it.

All of your answers above detail what your food selection habits are. Now discuss these results with your class and compare answers to get a bigger snapshot of food in your society and how family characteristics influence your choice of food.

4. Peer group

As children grow older, they are more influenced by what their peers eat as they share more meals together. This can be at school, sporting events, friend's houses or out at social events. Adolescents may find themselves being exposed to a bigger variety of foods through eating with friends. They may also find their own habits changing so they can fit in at school. For example, the need for belonging to a group may mean 13-year-old George throwing away the sardine and sweet dill pickle sandwiches mum prepared and buying the 'trendier' canteen food so he can fit in and not appear different to others in his lunch time social group.

Activity 10.4: Peer group scenarios

The situations described below are based on some adolescent food concerns. Read through them, then in small groups discuss and work out how they could be overcome.

1. **Customs:** your mother insists on giving you dhal (a well-seasoned lentil puree popular in Indian culture, usually curry-based) for your school lunch. While you find it delicious, you are getting some strange looks from others – especially when you take off the container's lid and the aroma goes everywhere, not to mention the comments you get when they see it. What can you do?

2. **Budget:** you love going to the movies and see it as a great chance to get out and socialise with some of the friends you grew up with. The problem is that the movies cost so much and you don't have enough money left for popcorn, choc bombs or to share a bite of pizza with your friends afterwards. How can this be solved?

3. **Your friends at school are really into their image:** the problem being they don't eat!! You need food constantly but don't want to be seen gorging your face while all the popular kids are around. But if you go without, you find you can't concentrate on school work and have no energy for athletics training after school. How can this be overcome?

5. Nutritional value

If consumers value health, they will purchase foods that are low in additives and have more nutritional benefit. Some may even take that further and purchase only organic certified produce to avoid chemicals found in foods. Health plays a vital role in the selection of food. Further, people who have been diagnosed with certain diseases or food intolerances may need a special diet.

Activity 10.5: Suitable foods for specific conditions

For the conditions listed below, outline what it is, what foods people with this condition need to choose and which ones they need to avoid.

- ADHD
- Anaemia
- Cardiovascular Disease
- Coeliac Disease
- Diabetes
- Lactose Intolerance
- Obesity
- Osteoporosis

Summarise your answer in a table such as the one presented below.

Condition	Description	Food concerns
ADHD		

All humans need nourishment (food and water) to survive, grow and function properly for day-to-day activities. Food is a basic necessity of all human beings. The type of food required depends on various factors related to the individual. The age of the person can determine different needs as younger children need a lot of energy foods as well as protein and calcium for growing bodies. Senior people would require less energy foods but still need a range of vitamins and minerals to ensure body functions are carried out normally.

Those who are involved in sport have specific food needs as well due to higher activity levels. A lot of carbohydrates for energy are needed, but athletes like gymnasts and jockeys would need to carefully monitor their weight levels, while boxers or those involved in contact sports like rugby and football can choose foods that have more kilojoules in them as weight is not necessarily as much of a concern.

Activity 10.6: Nutritional needs for specific age groups

Examine the following age groups and briefly discuss their specific nutritional needs.

- Toddlers
- Primary school students
- Adolescents
- Young adults
- Middle-aged adults
- Senior citizens

Environmental issues that arise from food and lifestyle choices

With the increased usage of processed foods due to our changing technology and lifestyle, there has inevitably been a great impact on the environment. While some changes have been positive, others have been negative.

Look at the different issues below and put a tick next to the positive changes and a cross next to the negative:

- ❑ increased air and water pollution from food processing
- ❑ increased use of pesticides and fungicides for the 'perfect' food product
- ❑ longer shelf life of foods
- ❑ more transportation of food used during processing
- ❑ added landfill due to packaging
- ❑ increased salinity due to land-clearing for farming practices
- ❑ soil degradation from farming
- ❑ increased range of foods available
- ❑ some seasonal foods available all year round.

1. Food availability

The concept of food and sustainable food supply has huge ramifications for the future of society. Whether food is eaten for pleasure or simply as a means of survival, changes in the way food is manufactured need to continue evolving to address wide-ranging world food issues like food shortage.

With increased knowledge and technology, trends and innovations develop that make use of these improvements, and this usually means that more varieties of food can be delivered safely to communities. This does mean that some issues arise from ensuring food is available in more usable formats.

- **The use of food preservation techniques** means that food can be available at different times of the year. Whether the food is frozen, canned, dried, irradiated or vacuum packed, consumers enjoy increased shelf life of products. Some of the environmental issues associated with this include:
  - increased packaging – use of natural resources
  - increased 'food miles'
  - storage space required
  - possible contamination through incorrect storage techniques
  - increased landfill from packaging waste.

Can you add to this list?

- **The availability of farming areas for growing foods** are at risk, especially in Perth as the urban sprawl extends to take over areas once used for agriculture. Farmers need to turn to more efficient growing techniques to be able to supply more food with less land. Broad acre farming techniques have changed due to less need for wool globally. Farmers are choosing between stock and crop, and many are only cropping their properties due to the damage and hassle associated with sheep on paddocks. If sheep are being stocked on the farm, many are bred for meat only and not the wool, due to market demand.

- With the increased need for farmers to produce more and more food for the growing population of the world, they have employed techniques like the **use of chemicals to improve the overall crop quality** and the use of **genetically modified seed** to combat disease and tolerance to growing conditions.

  Chemical use during the food production system is quite common in farming practices. While growing foods, chemicals are added in the form of artificial fertilisers and pesticides, all aimed to improve the crop's yield. Animals digest grasses, cereals and other food products which may have chemical substances in them, consequently storing them in their bodies and passing them on through the food chain.

  Manufacturers add chemicals to foods during food processing – these are known as **food additives**. This increases the availability of the food as many additives act as preservatives to improve the keeping quality of the food. These are all environmental concerns of which consumers need to be aware.

- Another environmental concern associated with the availability of food is the **water quality for both plants and animals**. The amount of rainfall is an issue, but so is the spread of salinity in farming districts throughout the South West and Upper Great Southern areas of Western Australia. Salt tolerant crops are being developed to address this issue and farmers are modifying farming techniques to reduce the impact of salt on their crops.

- **Introduced pests from imported products.** Australia is very lucky when it comes to pests and diseases. Because of the isolation of the continent, many pests that can destroy crops or bring disease to animals cannot physically get here. This is becoming a bigger concern, however, with increased importation of food and increased global travel by people. Quarantine controls restrict what foods/animals can be brought into Australia as a means of protection, but the system is not foolproof.

The cane toad was introduced to protect what crop of Australia? What is the ongoing issue of the cane toad's introduction?

2. 'Food miles'

This concept has been around for a while and was originally introduced in the United Kingdom. It is a measure of how far food has travelled from 'paddock to plate' with the notion that the less food miles the product has done, the better choice it is for the environment.

It has huge connotations for Australia because of the large size of our country. Any food freight, particularly by air and road (our most preferred option), consumes fuel and energy, releasing greenhouse emissions into the atmosphere, affecting the global climate. The more processed the food, it is assumed the more 'food miles' the product has when you examine all of the ingredients in the product.

Activity 10.8: Food miles

1. Examine a range of food products and determine their food miles from the information provided on the packaging.
2. Do you think the information on the packaging is a correct indicator of the product's food miles?
3. Is examining the concept of food miles by itself the best way to determine the energy needed to produce the food item? Explain your answer.

There has been considerable discourse over the concept of food miles. While staunch environmentalists support the notion of developing local and regional food supplies for sustainability, economists argue that trade is needed for sustaining Australia's economy and relationships with other countries. If we boycott their products, this could upset the trade balance and cause future issues over trade agreements. There is no doubt, however, that the further the food has to travel, the bigger the carbon footprint and effect on our planet. Consumers need to make their own choice.

Carbon footprint refers to the impact human activities have on the environment in reference to the amount of greenhouse gases produced, which is measured in units of carbon dioxide.

3. Packaging, recycling and waste

It is difficult to talk about the packaging of food products without examining the recycling issues. Most foods come to the consumer packaged in some way, with the exception of some fresh fruits, vegetables or various bread types that are placed in consumer's own shopping bags. It often depends on the shopping outlet and the preparedness of the consumer to provide their own shopping bags.

Packaging contributes to landfill and can be quite harmful to the environment. Consumers need to be aware of how and what can be recycled when making purchases to keep this waste to a minimum. Local councils are a great source of information on recycling and offer advice on a range of products.

Activity 10.9: Packaging

1. Examine some of the various types of packaging that are available in the market today. Complete the following chart, determining the effectiveness of these types of containers. Examine concepts like how it keeps food safe in terms of nutrient content and physical damage, ease of storage and environmental factors like how it is made and recyclable qualities. Summarise your findings in a table such as the one presented below.

Packaging	Advantages	Disadvantages	Interesting points
Plastic milk containers			
Cardboard milk containers			
Aluminium cans			
Tin cans			
Plastic shrink wrap			
Vacuum packaging			
Polystyrene trays			
Paper (like on butter)			
Crinkly plastic (like chip bags)			
Glass jars and bottles			

2. How do the values differ between consumers and manufacturers in regard to how goods are packaged?

3. Vacuum packaging is a relatively new technology where products are sealed in plastic after the partial removal of air from it. Research and describe its benefits to food manufacturers and consumers.

4. Many bags now have reusable seals, for example rice. Does this make storage easier? Why was this concept developed?

The following is a breakdown of some of the more commonly used products found in food packaging.

i. Plastic

Plastic is one of the more common substances found in food packaging and comes in a variety of forms. See Table 10.1 below.

TABLE 10.1: Plastics

Name of plastic	Description	PID*
Polyethylene terephthalate (PET)	A clear, rigid polymer, as used in soft drink bottles, fruit juice bottles and other rigid applications	1
High-density polyethylene (HDPE)	A conventional (not biodegradable) plastic, as used commonly in crinkly shopping bags. Also used in freezer bags, milk bottles and snack boxes	2
Unplasticised polyvinyl chloride (UPVC)	Hard, rigid polymer that can be clear. Used for clear cordial and juice bottles	3
Low-density polyethylene (LDPE)	A thick conventional (not biodegradable) plastic, as used commonly in more durable plastic carry bags and garbage bags. Also used for cheese wrapping, clear plastic gladwrap, fruit and vegetable bags and squeeze bottles	4
Polypropylene (PP)	Hard, flexible polymer. Used for bags containing potato crisps and biscuits	5
Polystyrene (PS) and Expanded polystyrene (EPS)	Foamed, light-weight polymer that is heat insulating. Used in meat and poultry trays, yoghurt and dairy containers and also vending cups	6

* Plastics Identification Code

The problem with these plastics is that they are non-biodegradable and need to be recycled via council kerbside collections to avoid being a hazard to the environment. Plastics that are recyclable are given a number in a simple coding system called the 'Plastics Identification Code'. This tells recyclers what type of plastic resin a product is made from. Be careful when putting out your recyclables and check with your local council what can be collected as it can change with different regions.

Did you know ...

Around 40% of PET plastic bottles produced are recycled which represents the single biggest volume of plastic type recovered.

Activity 10.10: Research plastics

Research how each of the above plastics can be recycled. For example, PET plastic can be remade into drink bottles, carpet fibres, clothing and detergent containers. What can the others be used for? You may be surprised!

This website – **www.repsa.org.au** – may help you. Click on the available downloads until you find the one on the plastics identification code.

ii. Cardboard and paper

This is another very common substance that is used in a wide range of food packaging in the form of boxes, cardboard tubes, sacks (eg. for potatoes) and bags. It may need some sort of treatment on it to repel moisture like grease-proofing or waxing. A huge advantage of cardboard and paper is that it is biodegradable and the addition of paper is encouraged on garden compost heaps. This substance is recyclable although cannot be reformed into 'new' paper or cardboard that will come into direct contact with food.

iii. Glass

Glass is found in food packaging in the form of bottles and jars. While these products can be reused in the home, glass is also a recyclable product collected by local councils. It is not biodegradable so consumers should make the effort to recycle it.

iv. Aluminium

Aluminium is found in beverage can bodies and is also used as foil for household wrap or in lining fruit juice cartons. It is very easily and economically recycled – in fact, aluminium containers can be recycled a number of times through local council collections.

Did you know ...

Aluminium formed from recycled cans takes only 5% of the energy required to produce aluminium from bauxite.

Activity 10.11: Packaging

Research the Australian Packaging Covenant and describe what it aims to achieve.

4. Food waste

Whether cooking at home, at school or within the food industry, there is a real concern for the amount of waste that is involved with food production. It makes sense to try and minimise rubbish produced as there would be both environmental and economic gains for the kitchen. If managed properly with waste-reducing measures, there would be less money being spent on food extras being thrown out and going to landfill during the food preparation process.

Waste from food preparation involves the trimmings from vegetables and fruits, bones and skin from meat, poultry and fish, and the shells from nuts. Other waste could include food that has gone 'off' due to poor kitchen management by either ineffective storage techniques or by lack of using the product on hand.

How to reduce waste in food production

The following are a few ideas of how to keep food waste to a minimum in the kitchen.

- Only purchase foods that you need – make a plan of the menu, verify quantities to be made, check stock availability and buy the right amounts of what you need.
- When purchasing, check use-by dates and ensure that you can use the product within that date.
- Use older products first – rotate stock on the shelf so that the older goods are at the front.
- Ensure all foods are transported effectively from the place of purchase and then stored correctly to avoid food spoilage.

Activity 10.12: Food waste

1. Can you think of any more hints to reduce food waste at home?
2. What do you do at school to ensure food waste is kept to a minimum?

Food scraps

As mentioned above, the trimmings from food vegetables and other scraps form a large amount of waste from the kitchen. Here are a few tips on what to do with these scraps:

- Make a stock from the bones and vegetable trimmings (clean ones).
- If suitable, use any leftovers as a means of garnishing.
- Add them to the compost heap for the garden.
- Add some of the scraps to a worm farm (remembering to omit any onion and citrus pieces).
- Keep some chickens in the back garden (if possible) – a great way to get rid of scraps and get fresh eggs at the same time.

Activity 10.13: Leftover ingredients (a)

In food preparation, we often have leftover ingredients. Explain what you could do with the following to ensure that they are not wasted. One example is slightly 'off' cream which can be used in cooking cakes like chocolate cupcakes.

- Old bread slices
- Leftover spaghetti sauce
- Excess rice
- Leftover meat from a lamb roast
- Slightly stale cake.

Can you think of any other examples? List them.

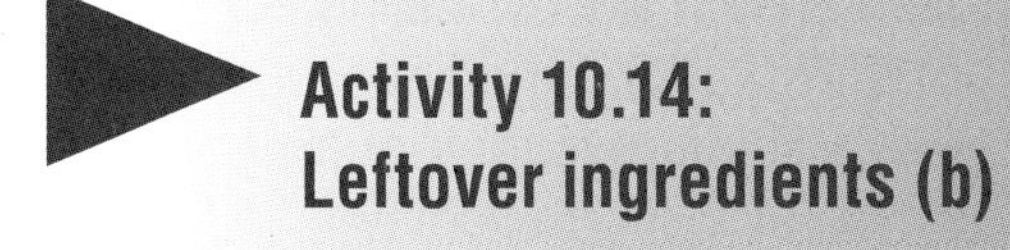

Activity 10.14: Leftover ingredients (b)

At certain times of the year, people with vegetables, fruit and citrus trees can have excess produce. What could you do/make with the following items if you have them in abundance?

- Lemons
- Apricots
- Strawberries
- Bananas
- Mangoes
- Tomatoes

Activity 10.15: Impact of food technology

There are many concerns that arise from increased food technology as we strive to feed a growing world population. Discuss, as a class, what are some of the things being done in the community to address this issue of waste. Should more be done?

Influences on adolescents' food choices

Media

The media can have a great influence on adolescents' eating habits. Whether it is from using forms of media like the computer or TV, viewing the ads that are screened before a movie at the cinema, listening to the radio or even looking at the simple billboards by the side of the road, adolescents are ripe targets for food advertisements. With so many images of perfection that are bombarded at them, they can be greatly affected and have a distorted view of what image they should have. This can lead to diet-related problems like bulimia, anorexia nervosa and obesity.

Marketing and advertising can influence food choices because it is a powerful tool that aims to work on the senses and give the consumer the notion that by using that product they will improve their quality of life. In the case of adolescents especially, this could be aimed at trying to achieve the following if they make use of their product:

- be cool
- 'fit in'
- be an individual
- look great
- excel at sport
- be successful at school
- attract the opposite sex
- be popular with peers
- be seen as a risk-taker
- be healthy
- improve body image.

This list is not all inclusive and successful advertising campaigns can transmit more than one aspect with each advertisement.

Activity 10.16: Advertising

Choose an advertisement for a food product aimed at adolescents from a newspaper, magazine or leaflet. Attach it to a piece of paper and describe the techniques used to get consumers interested in buying that particular product, drawing attention to specific aspects of the advertisement. What image are they trying to sell?

Techniques

Some large food companies spend a lot of money on sophisticated advertising campaigns using different methods to target their consumer market and attract their attention.

These advertising campaigns can use a combination of techniques like some of the examples listed below:

- **using catchy music** tunes and slogans
- **using celebrity, cartoon or identifiable characters** to represent the product – the cartoon figures may appeal to the younger adolescents while celebrities would attract the attention of older teenagers
- **citing qualified scientific research** that backs up any health claims the product may have including reference to genetically modified foods, nutrient content, recommended daily food intakes, the presence of gluten and the glycaemic index
- **using attractive body images** to gain attention and sell products
- **using specific fonts and/or graphics** that portray an image identifiable with the audience
- **using media practices** including colours/lines to represent the mood/image that is wanted to be portrayed (choose some basic colours and describe the emotions you have when thinking of them)

- **using language** that is recognised by the targeted audience as being particular to them
- **timing** the advertisements so that breakfast foods are presented in the morning and snack foods around lunchtime/afternoon in TV and radio media
- **using food styling** techniques to make food appear attractive

The aim of food styling is to make the food appear so delicious people either want to go out and buy it or make it if it is a recipe. When photographing food, stylists can go for the natural shot – straight out of the oven and captured on film perfectly – or they can utilise stylists' tricks. These tricks are usually done with undercooked food as the longer the product is cooked, the sooner it shrinks when away from the heat. Some techniques include using:

- eyebrow pencil to create lines in food to represent browning like char-grilled steak
- clear nail polish to brush over prawns to make them appear glossy
- cotton wool balls or tampons, soaked in water and microwaved to create the appearance of steam under items like vegetables, and in bowls of soup or porridge
- Scotchguard sprayed on pancakes to ensure syrup doesn't absorb into them.

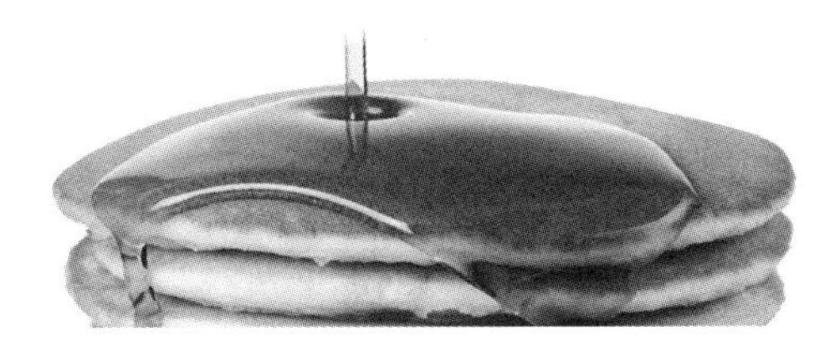

Activity 10.17: Techniques

Can you think of any other techniques? List them.

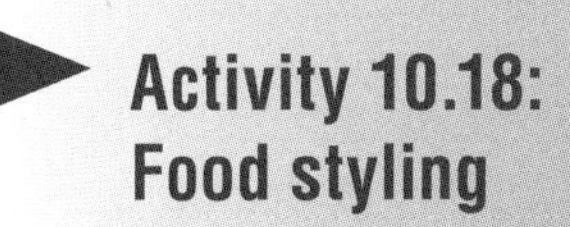

Activity 10.18: Food styling

What do you think you would use the following for when styling food?

- Tweezers –
- Paintbrushes –
- Sewing pins/toothpicks –
- Blowtorch –
- Vegetable oil and pastry brush –
- Nail scissors

Activity 10.19: Advertising techniques

1. Write examples of different advertisements you can think of for each technique described.
   - Use of celebrities
   - Catchy music
   - Cool body images
   - Use of identifiable colour
   - Specific fonts and/or graphics
   - Food styling techniques
2. Think of an advertisement for food that you have seen recently. What makes it so memorable for you? Describe the product, the advertising techniques used and the subliminal messages conveyed using the lists above.

Activity 10.20: Media influences

1. While advertising and marketing can be seen as a positive aspect since it informs the consumer of products available, it can have a negative influence on impressionable adolescents. How can you minimise the effects that the media has on their eating habits? Brainstorm ideas with others in your class.
2. In printed media, the use of airbrushing body images to make them 'perfect' needs to be regulated. Do you agree or disagree? Explain your reasons.

Ethical influences on food choices

Ethical issues refer to what is 'morally correct' and what are 'right and wrong' practices. Within the scope of food choices, consumers may base their selection on their beliefs and values. The following describes some of the ethical issues some people may face.

1. Animal welfare

Many consumers are sympathetic towards animals and do not like the thought of them being harmed for our food needs. This ranges from the dislike of killing animals for meat to any form of mistreatment in handling techniques. There has been public outcry about caged hens and the cruelty they face stuck in small enclosures for life. Producers are now marketing 'free range' eggs in response to the issue as well as free range poultry products. Recently, it has been announced that a series of free range pork production farms will be set up in the south-west of Western Australia to breed pigs for that market niche. This is in response to consumer demand for better conditions for the animals.

People involved with animal welfare are also continually debating over live trade export as semi-trailers full of sheep pass through Fremantle to the wharf ready to be transferred to the sheep carriers. Media have reported on cruelty to these animals at the hands of the overseas buyers, prompting animal activists to demand boycotts on sending the animals to those countries. It is a delicate balance between the welfare of animals and the need to keep farmers selling their produce.

Activity 10.21: Animal welfare

What would happen to farmers if the export trade would cease to happen because of the idea of cruelty to animals? What would happen to local food supplies?

For those consumers who are against animal cruelty, they can opt for a vegetarian diet. There are different types of vegetarians and the food industry recognises these individuals with most restaurants and food outlets carrying a range of menu options which do not include meat.

The scope of vegetarians ranges from those who choose not to eat red meat but will eat 'white flesh' like fish and chicken, to those who do not eat any animal product at all including fish, eggs and dairy products. Listed below are some types of vegetarians.

- Semi-vegetarians
- Lacto-vegetarians
- Lacto-ovo-vegetarians
- Ovo-vegetarians
- Vegans

Activity 10.22: Vegetarians

Research each type and find out what they can/cannot eat.

EXTENSION

Extension Activities

Those listed above are the more common types of vegetarians but there are also two more that are known as 'macrobiotic' and 'fruitarians'. What is so special about these particular vegetarian diets?

2. Fair trade

Fair trade is known as an alternative approach to normal trade practices and is based on a partnership between the producer and the consumer.

> **Fair trade**... 'is a trading partnership based on dialogue, transparency and respect that seeks greater equity in international trade. It contributes to sustainable development by offering better trading conditions to, and securing the rights of, marginalized producers and workers – especially in developing countries.'
>
> Fairtrade.net

This fair trade agreement allows workers the opportunity to improve their lives and plan for the future. Fair trade offers the ethical consumer a powerful way to reduce poverty through their everyday shopping by purchasing Fair Trade products. When a product has the FAIRTRADE Mark, it means the producers and traders have met Fair Trade standards that address the imbalance of power in trading relationships, unstable markets and the perceived injustices of conventional trade.

Activity 10.23: Fair trade

What are some examples of Fair Trade products? List them.

3. Resource use

Ethical consumers who are concerned with resource use try to ensure that all of their food activities are completed in a sustainable manner. This could include:

- purchasing products that are in recyclable containers
- re-using food containers
- using own containers – eg. own coffee cup at café for take-away coffee
- making sure wastage is minimal – using food up before the use-by date expires
- only purchasing what is needed.

Activity 10.24: Resource use

Can you think of any other methods?

With the understanding that earth has finite resources, ethical consumers would be ensuring that they use them as wisely as possible.

4. Country of origin

Consumers worried about food miles or politics or unfair trading practices may check the country of origin when choosing food products. With the understanding that 'buy local' is preferred, consumers in Western Australia would be looking for the West Australian symbol that indicates the product is presumably grown and produced here. Labelling must be checked properly though, as even though it is made in Western Australia, it could contain imported goods.

Activity 10.24: Country of origin

Check the pantry at home, examine the labels and make a list of all the Western Australian food products you can find. Are you surprised at the amount you found or did not find?

Health issues that arise from food choices

It is no secret that there are concerns for the health of many Australians with cases of obesity, heart disease, cancer and diabetes on the rise, to name just a few of the common health problems. This is not only linked to an ageing population, but also to younger generations where children and adolescents are now displaying health issues related to being overweight.

The increased availability of ready-made food and our sedentary or inactive lifestyle (compared to previous generations) have helped contribute to many of these concerns. Let's look at some of the health issues that arise from inappropriate food choices.

1. Malnutrition

This is a lack of nutrition that can be caused by not having enough to eat, not eating enough of the right foods or the inability to absorb or metabolise the nutrients. This is an issue that many do not associate with Australia, but it does exist in all areas of the population. The following groups are particularly at risk:

- the elderly
- the Indigenous population
- low income earners or homeless people
- ESL families
- people with physical disabilities
- those with laack of nutritional knowledge
- those with health issues that prevent the body from absorbing nutrients, eg. Crohn's disease
- people with drug or alcohol problems.

Activity 10.25: Malnutrition

For each of the groups above, describe how they can be malnourished. The elderly have been done for you as an example.

Elderly – a lot of effort needed to make meals so many may not bother, perhaps have poor-fitting dentures that do not make eating easy, may have dementia and forget to eat, may not be able to get to the shops easily or have lack of mobility to be able to move around and prepare meals.

Malnutrition is a serious issue that occurs when a person's diet does not contain the correct balance of nutrients to meet the demands of their body for normal functioning. That can affect a person's mood, behaviour, growth, physical health and their body functions, all of which have an impact on the body's overall health.

Activity 10.26: Resource use

What can be done to ensure these groups of people do not suffer from malnutrition?

What would you do if you suspected an elderly relative was malnourished?

2. Underweight

Being underweight means weighing less than the normal amount for one's age, height and build. It is where the body weight is considered too low to be healthy. One way to check this is by using **Body Mass Index (BMI)**, which is the relationship between your weight and your height. It is a useful tool to determine your body's health as BMI is used to estimate your total amount of body fat. It is calculated by dividing your weight in kilograms by your height in metres squared (m^2). The following is a breakdown of the BMI ranges as determined by the World Health Organisation.

If your BMI is:

- **under 18.5:** you are very underweight and possibly malnourished
- **18.5 to 24.9:** you have a healthy weight range for young and middle-aged adults
- **25.0 to 29.9:** you are overweight
- **over 30:** you are obese.

There are many different BMI calculators that are online and can do the sums for you. The Heart Foundation has one that is easy to use and understand. Just use the following link:

www.heartfoundation.org.au/healthy-eating/Pages/bmi-calculator.aspx

Activity 10.27: BMI calculators

BMI calculations are designed for men and women over the age of 18. Why should younger people not use it as a guideline?

Using the BMI indicator, if a person's BMI is under 18.5, they are considered underweight. Being underweight is an issue as it could have been caused from the following factors:

- stress
- anorexia nervosa
- bulimia
- malnutrition
- disease or physical injury/impairment.

Some of these causes are from mental issues which need professional help before they worsen. If a person is underweight, they are at more risk of developing:

- compromised immune function
- respiratory disease
- digestive diseases
- cancer
- osteoporosis.

Activity 10.28: Bulimia and Anorexia

Research the difference between Bulimia and Anorexia Nervosa. Why are these conditions so difficult to treat?

3. Overweight

This is a major issue in Australia today. According to the Australian Institute of Health and Welfare (AIHW), rates of overweight and obese people are continuing to rise. Consider the following statistics from the Australian Bureau of Statistics 2011-12 analysis.

- 70% of males and 56 % of women aged 18 years and over were overweight or obese.
- This rate has increased 5 and 6 percentage points from the 1995 results. The results are BMI based and have been age standardised to account for differences in the population structure over time.
- 24%of boys and 27% of girls were either overweight or obese according to measured BMI.
- 7% of boys in the 5-12 age group were obese, an increase of 3 percentage points from 1995.
- 18% of girls in the 13-17 age group were overweight, which is up 6 percentage points from 1995.

Activity 10.29: Overweight

Why are these trends occurring? Come up with a list of reasons.

Do the reasons differ for the various age groups?

People who are overweight or obese may be significantly affected in their health, social and economic outlook. Lack of exercise and diet is a usual cause of this condition, although genetics can have an influence. Being overweight or obese can lead to health issues like:

- coronary heart disease
- Type 2 diabetes
- some cancers
- joint problems like knees and hips
- sleep apnoea.

Besides the effect of obesity on a person's health, it is also an economic problem. In 2008, it was estimated that the total annual cost to Australians, including loss of productivity and carers' costs, was about $58 billion.

This is one of the biggest problems facing Australian society today. What is being done about it? Should the government do more? What more can happen to reduce this significant problem?

4. Allergies/intolerances

There is also the growing concern of allergies and food intolerances, the incidences of which seem to be increasing. Examples include allergies to peanuts, tree nuts (walnuts, almonds, cashews and so on), cow's milk, eggs and soy, wheat and seafood proteins. The food allergy can affect different parts of the body including the mouth, throat, lungs, skin and gastrointestinal tract.

With so much processed food in our diets, consumers need to read packaging labels carefully so as to avoid foods they may react to. The same machinery that made peanut biscuits may also be used to later make plain ones, but residue left on the equipment could still transfer to the next food product and cause a reaction. That is why many labels, for example ice cream, have the warning 'may contain traces of nuts'.

Did you know ...

Food allergies or intolerances can cause anaphylaxis or anaphylactic shock which is a severe, sudden and potentially life-threatening reaction to particular foods.

Activity 10.30: Food allergies

1. What is meant by 'Nut-Friendly Schools'? Research and explain, providing examples of guidelines used in the initiative.
2. Describe what is meant by 'gluten intolerance' (otherwise known as coeliac disease) and how people who have this can still eat relatively normal foods.
3. Either go to a supermarket or go online and examine some of the gluten free foods available. List them, and then compare their cost with 'normal' products. Study the comparisons and then describe whether you think this price variance (if any) is justified.

Useful Websites

www.accc.gov.au/consumers/groceries/grocery-unit-prices: Contains details about unit pricing, how it works and examples

www.packagingcovenant.org.au/: Describes the national packaging covenant and what it aims to achieve

www.abs.gov.au/ausstats: Contains information about obesity trends and the 2011-12 data analysis

www.aihw.gov.au/overweight-and-obesity: Obesity and overweight data

www.health.gov.au/internet/main/publishing.nsf/Content/health-pubhlth-strateg-hlthwt-obesity.htm: Obesity and overweight data

CHAPTER 11
Laws and Regulatory Codes

Key Concepts

- Workplace regulations for safety and health
- Safe food handling practices
- Labelling requirements for food and beverage products available in Australia

WORKPLACE PROCEDURES FOR SAFETY AND HEALTH

It is important for safe work techniques to be upheld both in the classroom and in food outlets. This relates to both personal safety for the food worker and food safety where food is kept free from contamination. Both hygiene and safety have been discussed in previous chapters, but this chapter will extend that basic knowledge and also investigate the laws and regulations that ensure a safe, local supply of food.

Protective clothing and footwear

It is important that protective clothing and footwear are worn correctly in the food industry to ensure high standards of food safety for the worker and consumer.

Because of the hazardous nature of food preparation involving working with hot surfaces and foods, a proper uniform should be worn to protect the worker. The uniform should extend from head to toe, to ensure the worker is protected from spills and splatters touching the skin directly. The following is a description of some uniform items.

- Hat: usually fabric, but can be paper, and hairnet used for long hair.
- Beard net: some men with long beards will need to wear a net to enclose the facial hair.
- Necktie: knotted around the neck to absorb perspiration.
- Jacket: white, long-sleeved and double breasted with raised buttons so that the jacket can be easily and quickly removed in an emergency. (The double layers of the jacket helps protect from heat and scalds.)

- Trousers: usually black and white checked with drawstrings and no cuffs as liquids could become trapped there if spilled.
- Apron: waist ¾ length below the knees – the 'bib' variety can get caught in machinery so should be avoided. It should be properly tied at the back or at the front if ties are quite long.
- Shoes: comfortable closed-in leather lace-ups (not sneakers) with non-slip soles.
- Fabric used: because workers are in a hot environment, clothing should be able to 'breathe', so natural, strong and durable fibres like cotton should be used. The colour white is preferred to give an image and feeling of cleanliness. Fabric should be smooth and one that does not shed fibres. The garment should be designed with low-flammability features, for example close-fitting sleeves.

Activity 11.1: Protective gear

Can you think of any other protective gear that needs to be worn by workers in different food environments? (Think of butcher's as an example.)

Activity 11.2: Hats and food handlers

Use the following website:
www.public.health.wa.gov.au/cproot/1558/2/Hats_and_Food_Handlers.pdf
or visit the Public Health WA website and type in 'hats and food handlers' and then answer the following questions.

1. Why wear a hat when handling food?
2. When should the hat be worn?
3. When should hair be fully covered and when can you only partially cover it?
4. When is a hat not required when working with food?
5. What happens if a food worker chooses not to wear a hat in the food service industry?

Personal hygiene

Food workers need to ensure that they follow the appropriate hygiene standards to ensure they do not contaminate food products and carry this risk through to the consumer. A high standard of personal cleanliness must be maintained at all times.

- Hair: needs to be washed regularly and you need to address any problems with dandruff. Tie hair back or wear a hair net to prevent loose strands falling into the food and contaminating it. Do not touch your hair while cooking and never brush it in a food preparation area. Facial hair needs to be kept to a minimum.

- Jewellery: needs to be kept to a minimum although a wristwatch can be useful. Wear plain ring bands only and no bracelets or dangly earrings. Facial piercings are also not encouraged – they should be removed or covered.
- Uniform: must be cleaned every day.
- Skin disorders: like acne should be treated but do not apply lotions or make-up in the food preparation area. For conditions like eczema, psoriasis and dermatitis, make sure skin is well covered and wear protective gloves where necessary. Any open wounds or sores need to be covered with a coloured, waterproof dressing with disposable gloves to be used if necessary.
- Ears: bacteria enjoy the warm, moist environment of ears so keep fingers well away from them during food preparation.
- Mouth: germs in saliva live in the mouth so avoid touching it or your lips when preparing food. Use a separate clean spoon to taste food for seasoning and wash immediately after use, no double-dipping. Never eat while preparing food, only taste it with a clean spoon – and no licking of fingers or double-dipping!
- Nose: wipe your nose with a clean tissue then discard it after use immediately and wash hands. When sneezing, turn your head away from the food and cover your nose with a tissue or hands to prevent bacteria spreading, then discard the tissue and again wash hands thoroughly.
- Body odour: should be kept to a minimum and avoid overusing body sprays and perfumes. Shower or bath at least once a day to maintain a good level of personal hygiene.
- Feet: wear good clean absorbent socks (cotton is ideal) to absorb perspiration. Make sure shoes are comfortable and not too tight fitting.
- Hands: need to be washed thoroughly with a vigorous action in soapy water or an antibacterial solution as hot as your hands can take it for 15 to 20 seconds, and then rinsed under clean, running water. They need to be either dried under a hot air dryer or on individual, disposable paper towels. Remember that bacteria love warm, moist environments so ensure hands are dried thoroughly. Fingernails should be clean, trimmed short and free of varnish.
- No sitting on workbenches or tables, no spitting on/near food preparation areas, no smoking in the food premises.
- Do not cook if you have a contagious condition like skin infections, boils, infected sores, dysentery, infectious hepatitis, salmonella infections or the simple cold or flu – germs are easily transmitted to the food.

Remember: although it is not law, there are guidelines in place that specify the minimum requirements for a return to work for food handlers after an illness like gastroenteritis. This protects the community against contracting contagious illnesses from contaminated workers, and these recommendations should be complied with.

Activity 11.3: Personal hygiene

1. Which of the above dot points do not necessarily apply to you during food preparation lessons in your classroom? Explain your answers.
2. If you have a cold, what precautions do you need to take to ensure you do not pass on your illness to others when preparing food for them?

Activity 11.4: Physical safety and hygiene

1. Using the picture below as a guide, list all the things a person in the food industry should do to ensure a high standard of personal cleanliness and physical safety. Be specific with responses, describing what is needed.

Physical safety		Hygiene
•		•
•		•
•		•
•		•
•		•
•		•
•		•
•		•
•		•

2. What rules do you have in your foods class at school that are concerned with safe work practices in regard to your personal presentation?
3. What procedures do you follow when cooking food at school that ensure safe food hygiene practices?
4. When do hands require washing during food preparation and serving? List as many instances as you can that you consider important to ensure food safety.
5. What advice would you give a food worker about body odour and the use of perfumes/lotions?

Emergency procedures

In the food industry, there needs to be plans in place to be able to deal with emergency situations. In order to be prepared, the areas of risk need to be identified.

Table 11.1 presents a summary of typical hazards associated within the food industry.

TABLE 11.1: Typical hazards in the food industry

Hazard	Potential harm or danger
Manual tasks	Overexertion can cause muscular strain.
Electricity	Working with electrical equipment could cause injuries from exposure to live electrical wires including shock, burns and cardiac arrest.
Slippery surfaces	Slips, trips and falls can cause fractures, bruises, lacerations, dislocations and concussion.
Machinery and equipment	Body parts being caught by moving parts of machinery can cause fractures, amputation, bruises, lacerations and dislocations.
Cutting equipment	Slipping with sharp cutting utensils can cause cuts, lacerations, even amputations.
Hazardous chemicals	Toxic or corrosive chemicals may be inhaled, come into contact with skin or eyes causing poisoning, chemical burns and irritation.
Hot surfaces	Hot surfaces and materials can cause burns. Exposure to heat can cause heat stress and fatigue.
Steam burns	Burns caused by exposure to moist heat such as hot liquid or steam.

How to deal with simple first aid procedures has been dealt with in a previous chapter. First aid kits should be accessible at each workplace and provide basic equipment for administering first aid for injuries such as:

- cuts, scratches, punctures, grazes and splinters
- muscular sprains and strains
- minor burns
- amputations and/or major bleeding wounds
- broken bones
- eye injuries
- shock.

The kit should be easily accessible with well-placed signs indicating its location.

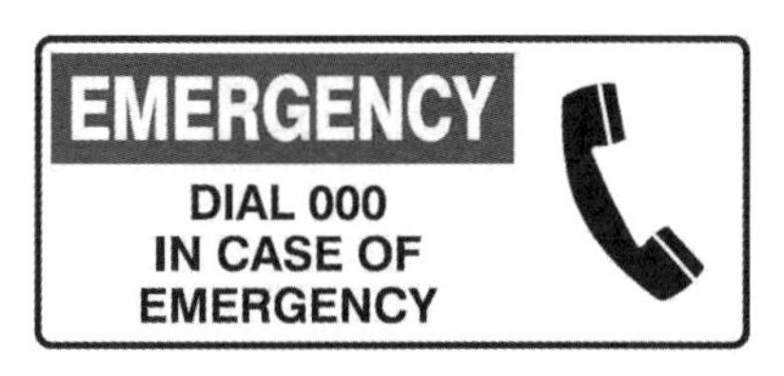

Remember that with care and attention to the task at hand and knowing the risks of what can happen most accidents in the food preparation area can be avoided. There should be a person trained in first aid in each workplace, and emergency phone numbers should be prominently displayed. For example, 'In case of emergency dial 000'.

Other emergencies could involve fire, gas leaks or threats of violence. In the first two instances evacuation procedures should be followed, which includes the activation of isolation switches that disconnect gas and electricity supplies. Workers and patrons should follow evacuation signage and move to an approved assembly area away from danger.

Activity 11.5: Emergencies

For each of the following scenarios, describe how to cope with the situation safely.

1. An oven mitt catches alight when accidentally placed on an open flame.
2. Hot oil in a saucepan overheats and catches alight.
3. You notice a strange smell coming out from an electric beater.

Activity 11.6: Safety in your workplace area

1. Where are the safety stop switches, fire extinguisher and fire blankets in your room?
2. Familiarise yourself with your classroom's safety procedures and also take note of the evacuation process needed for a large scale emergency. Where is the meeting point for your class if the room has to be evacuated? Write it down and detail the procedures that need to be followed if there is an evacuation.
3. Describe how to deal with the following, writing down the steps to be followed in each situation with guidance from your teacher:
   - deep cut to the finger while chopping meat products
   - scalded forearm from lifting the lid on a large stockpot filled with boiling soup.
   - unattended frypan where oil overheats and catches on fire
   - tea towel left near a gas flame catches alight
   - faulty power cord plugged into electrical outlet short circuits causing a loud noise and sparks.
4. How can each of the scenarios above be prevented? Discuss with the class and detail the safety measures to be taken to ensure they do not happen.

Signage for procedures and hazards

If there is a hazard identified in the workplace, signs or information need to be in place to warn of the danger or potential danger. You would already have seen a few examples in the previous pages on 'Emergency Procedures'.

If there are slippery floors, signs need to be erected near or over the area to warn of the danger, especially if a liquid has recently spilt there.

Signs also need to be erected for certain procedures. Exits for emergencies must be clearly and prominently displayed.

Warnings about smoking should always be displayed where relevant. Most food establishments discourage smoking inside as it affects the pleasurable experience of dining. It is against the law in Western Australia to smoke in indoor areas of clubs, bars and pubs. It is also an offence to smoke in outdoor eating areas like cafés, lunch bars, hotels, restaurants and delis. This sign should clearly warn patrons about their smoking rights.

Activity 11.7: Signs

Can you find and draw in signs for the following?

- Washing Hands
- Chemical Hazard
- Fire Extinguisher
- Safety Blanket
- Fire Hose
- Electrical Hazard

Safety data sheets (SDSs) provide information about any hazardous chemicals that are handled, used or stored at any workplace. The SDS gives information about the identity of the chemical, possible health effects, safe handling and storage procedures, first aid requirements and disposal considerations. Manufacturers, importers and suppliers of hazardous chemicals have a duty under the Work, Health and Safety Regulations to ensure that the current SDS is provided to a person at the workplace if the person asks for it. SDSs were previously known as Material Safety Data Sheets, but have been rebranded to align with international standards.

Activity 11.8: SDSs

Do you have any Safety data sheets in your foods room? Find out!

Safe posture, including lifting, bending and standing

All workers need to be aware of safe posture techniques to ensure they do not tire easily on the job or give themselves an injury. The following are general guidelines for working in the food preparation area.

Standing

It is important to have good posture when standing for extended periods of time as an incorrect stance can led to muscle strain. Try to remember:

- Keep your shoulders straight and head high.
- Keep your weight evenly balanced on both feet and avoid leaning on surfaces.
- Wear comfortable, non-slip shoes.
- If you are preparing food, try to choose a work surface which is the correct height for you. It should be one where you do not have to bend over, but instead have the elbows slightly bent with back straight when cutting food on a chopping board.

Lifting and bending

Although you are usually working on low-scale production in your foods class, sometimes it is necessary to lift heavy weights. Whether you are carrying platters of food, a stack of baking trays or boxes filled with food products, care needs to be taken to ensure no injuries are sustained – especially with your back. Remember the following points.

- Bend using your legs – do not bend at the waist.
- Lift using your leg muscles, not your back muscles.
- Only lift what you can carry comfortably at one time.
- Keep the load close to your spine.
- Turn by moving your feet; do not twist your back.
- Make sure your pathway is clear of obstructions.
- Ensure you can see where you are going and the load does not obstruct your line of sight.
- Have someone assist to open doors and check for obstructions.

Laws and regulatory codes

The following laws/codes are applied to business owners, workers and clients in the food industry in relation to workplace safety and health.

- Occupational Health and Safety (OHS) Act: passed December 1984, this assists workers and business owners by providing guidelines to follow to manage safety in the workplace. WorkSafe is the government agency in Western Australia that is part of the Department of Commerce that administers the Act. WorkSafe aims to reduce workplace fatalities and injury and disease rates in accordance with national targets.

More information can be found at www.commerce.wa.gov.au/Worksafe/

In other states of Australia, new Work Health and Safety (WHS) laws commenced on 1 January 2012 to merge together the previous OHS laws across Australia. The model WHS Act is not significantly different to OHS laws but makes it easier to for businesses and workers to comply with their requirements across different states and territories. Only Victoria, Tasmania and Western Australia are not totally using the WHS model at this stage.

More information can be found at www.safeworkaustralia.gov.au

Activity 11.9: OHS

Why is it necessary for the government to introduce legislation about workplace safety?

Occupational Health and Safety officers are people who coordinate health and safety systems within the workplace. They promote OHS and provide advice on how to prevent accidents from occurring as well as offering information on other health issues related to that occupation or work place. They can also:

- ensure employees are using protective gear such as safety footwear and protective headgear as determined by regulations
- identify and trial work areas for potential accident and/or health hazards and implement appropriate control measures
- check to make sure that dangerous materials are correctly stored
- conduct training sessions for management, supervisors and workers on health and safety practices and legislation
- develop occupational health and safety systems, including policies, procedures and manuals.

There are also other tasks that they can do that are not specifically related to the food environment.

Activity 11.10: OHS officers

Using the above points, imagine you are an OHS officer inspecting the premises at your school where you do Food Science and Technology.

Design a checklist of what you need to look for to ensure OHS guidelines are met at your premises.

SAFE FOOD HANDLING PRACTICES

If you decide to go to your local shop and purchase food, how do you know that what you are paying for is safe to eat and free from contamination?

Food poisoning results, on average, in 120 deaths, 1.2 million visits to doctors, 300,000 prescriptions for antibiotics, and 2.1 million days of lost work each year. The estimated annual cost of food poisoning in Australia is $1.25 billion (Food Safety Information Council www.foodsafety.asn.au). The council also estimates that there are 5.4 million cases of food-borne illnesses in Australia each year. It is the responsibility of food businesses to provide food safe to eat to consumers and this can be achieved by following set standards and procedures.

The Western Australian state and local governments closely monitor the supply of foods through laws and regulations. The Department of Health has a Food Safety Division whose role is to protect public health and safety by ensuring food that is available for sale in Western Australia is safe for human consumption. Environment health officers within local governments enforce the relevant food laws and legislation in their administrative area. They regularly inspect premises where food is prepared for commercial purposes to check that the procedures and hygiene standards comply with relevant regulations. The health officers also take food samples for microbiological and chemical testing, which they send to a pathology laboratory to check for food contaminants. They also check on restaurant staff and if they observe the staff handling food in an unhygienic manner, an environmental health officer will educate the operators about the right food-handling procedures and conduct ongoing, regular inspections to ensure that the problem or behaviour is corrected.

The following section outlines areas where workers in the food industry need to be particularly careful in relation to food safety.

Safe storage and thawing of raw and processed foods

When OHS officers check on establishments to ensure that proper techniques are used to ensure food safety, they closely monitor how the food is prepared and how it is stored. Poor handling and storage techniques can be blamed for most food-borne illnesses and it is vital that people dealing with food products know how to keep their food safe.

There are three main types of storage. These are:

1. Dry storage for foods such as flour, rice, spices and canned foods. The storeroom must be cool and well ventilated with low humidity. Products like onions, potatoes and garlic need dark areas for storage otherwise they can sprout shoots.

2. Cold storage for perishable foods such as meat, dairy, fish and poultry. Foods must be stored covered with plastic film or a damp, clean cloth, or sealed in containers to prevent food spoilage.

   Only mushrooms need storing in paper bags.

3. A freezer for foods that require longer storage or for keeping foods frozen such as ice cream. Most foods freeze successfully and frozen foods should be clearly labelled with what it is, the quantity plus the date it was frozen.

Temperature danger zone

Most bacteria will grow between 5° and 60°C therefore it is known as the 'temperature danger zone' as bacteria can grow to unsafe levels when food is left in that temperature range.

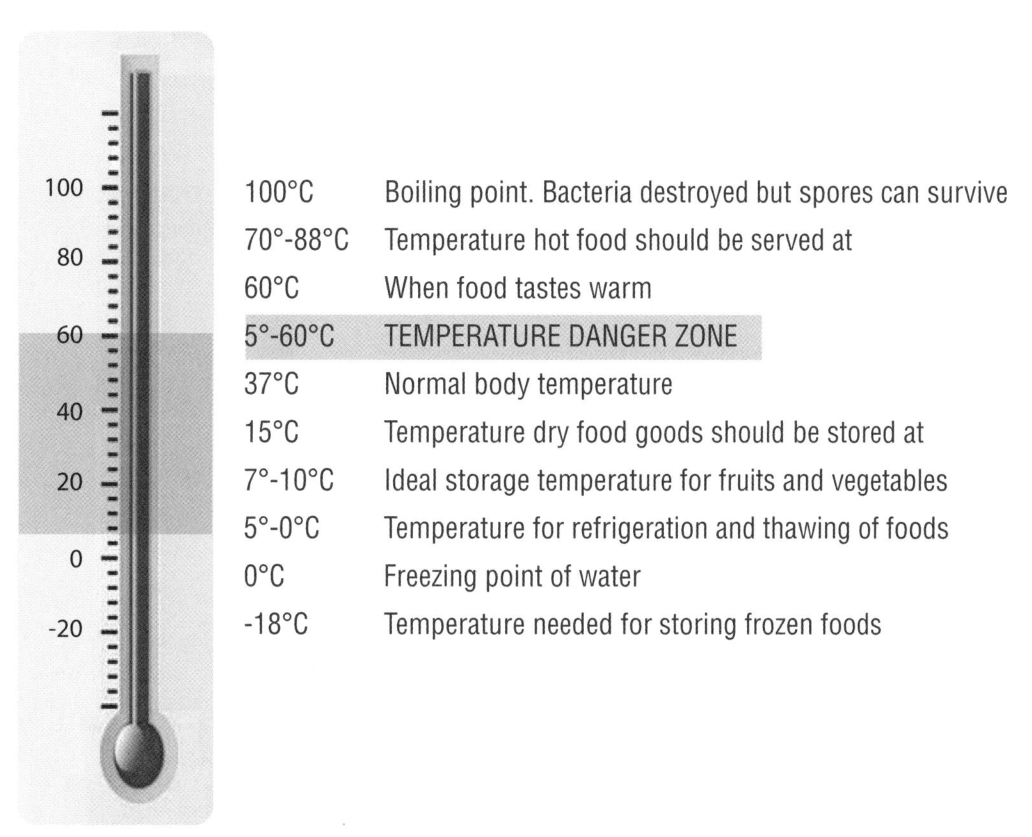

100°C	Boiling point. Bacteria destroyed but spores can survive
70°-88°C	Temperature hot food should be served at
60°C	When food tastes warm
5°-60°C	TEMPERATURE DANGER ZONE
37°C	Normal body temperature
15°C	Temperature dry food goods should be stored at
7°-10°C	Ideal storage temperature for fruits and vegetables
5°-0°C	Temperature for refrigeration and thawing of foods
0°C	Freezing point of water
-18°C	Temperature needed for storing frozen foods

Foods that are at risk include:

- raw meats, cooked meats and food containing meat, such as casseroles, curries, lasagne and meat pies
- dairy products and foods containing dairy products (milk, cream, custard) and dairy-based desserts

- seafood (excluding live seafood) and food containing seafood such as seafood salad
- processed fruits and vegetables such as cooked vegetables and ready-to-eat fruit packs unless vacuum sealed
- cooked rice and pasta
- vegetables, garlic or fresh herbs in oil where a food acid does not appear on the ingredient list
- gravies and sauces
- sandwiches and salads that contain the above foods.

Thawing foods safely

When food is frozen it stops bacteria from growing but it does not kill the bacteria – it lies dormant. This is why you always need to defrost food correctly so that the bacteria do not continue to grow and cause food poisoning.

There are several guidelines to follow when defrosting food:

- Always defrost the food thoroughly before reheating or cooking.
- Use a microwave or the fridge to defrost food – never leave it out on the bench. If using the fridge, make sure the food is contained and does not drip onto other foods.

- If food is defrosted in the microwave, it needs reheating or cooking immediately.
- Never refreeze defrosted foods!
- Allow ample time for foods to thoroughly defrost – especially for large dense items like meat for roasting or Christmas puddings.
- Always reheat cooked, defrosted foods to 75°C or hotter.

Remember these points to protect your food

- Always store hot foods above 60°C and cold foods below 5°C.
- When reheating food, heat right through until steaming hot.
- Cook food thoroughly.
- Cool food rapidly. If the volume is greater than 3 L, place the food in small containers to speed cooling.
- Store cooked and raw food separately.
- Keep high-risk food out of the temperature danger zone.

Prevention of cross-contamination

Cross-contamination occurs when bacteria and viruses are transferred from a contaminated surface to a non-contaminated one. The bacteria and viruses can come from people, work surfaces or equipment, and other foods. An example is when bacteria from the surface of raw foods like meat or poultry come into contact with food that's ready to eat like salads fruit or cooked meats. This can also occur with the dirt from unwashed vegetables like potatoes. The bacteria on the raw food are killed when the food undergoes the cooking process, but the bacteria on the ready-to-eat foods remain, possibly multiplying.

The transfer of bacteria from raw to ready-to-eat foods can occur through dirt on hands or cooking equipment. Chopping boards, plates and knives that have been in contact with raw food need to be carefully washed with hot water and detergent, then rinsed and thoroughly dried before being used for ready-to-eat foods.

How to prevent cross-contamination

- Store foods correctly, sealed tightly with lids in the fridge/freezer and covered, off the floor for other items.
- Keep raw foods away from ready-to-eat foods.
- Use separate boards for meat, breads, fruits/ vegetables, dairy and seafood.
- Wash equipment thoroughly in hot soapy water before using on another food or use separate equipment.
- Wash hands thoroughly with hot soapy water when working with cooked/uncooked foods.
- Avoid any unnecessary contact with food.
- Keep any cleaning materials/chemicals and other non-food items away from the food.
- Regularly change or wash and sanitise any cleaning cloths.
- If possible, air dry equipment and utensils after washing and sanitising.
- Always wash and thoroughly dry hands when starting food preparation, when changing tasks or returning from a break.

Activity 11.11: Cross-contamination situations

Explain how cross-contamination can occur in each of the situations listed below and give an example.

- Storage
- Chopping boards
- Food preparation
- Hands
- Sink
- Dish cloths and tea towels
- Garnishes
- Serving of food

Activity 11.12: Cross-contamination

For your next cooking lesson, examine the recipe and work out all the instances where cross-contamination may occur. Compare your answers with others and determine how you can prevent this from happening.

Clean equipment and work surfaces

Ensuring all equipment and surfaces are free from bacteria is a vital step in providing safe food to eat. Although it is not a legal requirement to clean and sanitise at specified times and temperatures, food establishments must ensure their equipment and work surfaces are cleaned regularly. The following descriptions outline the differences between cleaning, sanitising and sterilising in the food industry.

- Cleaning: a process that removes visible contaminants like food waste, grease and dirt from surfaces with the use of water and detergent. This process removes micro-organisms but does not destroy them nor is intended to.
- Sanitising: destroys micro-organisms with the use of both heat and water, or by chemicals. It is designed to reduce the numbers of micro-organisms present on a surface.
- Sterilising: a procedure designed to destroy all micro-organisms and spores. This process is not usually carried out with eating and drinking utensils or food contact surfaces.

Cleaning and sanitising needs to be done as separate processes because if a surface is not cleaned properly, the sanitiser's effect is diminished due to food residues and detergent that inhibit its effectiveness.

Any business working with food must use a cleaning process that ensures the utensil or food contact surface looks, feels and smells clean. Thorough cleaning can be achieved by:

- pre-scraping the utensil or surface to remove most of the food residue present
- using warm to hot water (54°C-60°C), detergent and agitation to remove food residue
- rinsing the detergent and food residue away.

Activity 11.13: Cleaning

What special cleaning techniques do you employ at school that are different to those used at home?

Activity 11.14: Instructional safety posters

Using computer technology, you are to work in small groups to create a series of posters or small pamphlets instructing others on the potential hazards that exist within your Food Technology classroom.

Outline recommendations for health and safety procedures as well as hygienic practices for lower school students using these similar facilities.

FOOD LAWS ASSOCIATED WITH FOOD SAFETY

All food laws within Western Australia are governed by a stand-alone piece of legislation The Food Act 2008. It presents a 'paddock to plate' approach to food regulation and takes over from the Health Act of 1911. The Act works to promote national consistency between Australian states and territories by incorporating the Food Regulation Agreement of 2000 (updated 2008) and also the Australia New Zealand Food Standards Code (FSC) in its entirety. The FSC is managed by Food Safety Standards – a bi-national agency that develops and administers the code which lists requirements for foods such as additives, food safety, labelling and GM foods.

Labelling requirements for food and beverage products available in Australia

Food labels are important. They provide the consumer with a wide range of information that helps them make food choices. Whether it is for inspecting use-by dates, or what ingredients are listed to check for allergens or determining country of origin, there are certain requirements for the types of information that must appear on food labels.

In Australia, there are laws that prescribe what is required when labelling food. The organisation primarily responsible for this is the Food Standards Australia New Zealand (FSANZ) which has several roles. It:

- monitors aspects of food laws such as the Food Standards Code and Food Labelling
- provides standards for the processing and primary production in a 'whole food chain' approach
- conducts national food surveillance to monitor the food supply in Australia.

Although this is covered by federal legislation, it protects the local as well as national food supplies in ensuring foods are not labelled in a misleading manner.

Labelling of foods

When consumers are deciding on what to purchase in the supermarket, the product label is often the last form of advertising or persuasion the food company has to get you to buy their particular product. The food laws and fair trading laws in Australia (and New Zealand) determine that labels do not misinform the consumer through false, misleading or deceptive representations. In Australia, this legislation includes the Australian Consumer Law (ACL) contained in the Competition and Consumer Act 2010, and State and Territory Fair Trading Acts and Food Acts which are enforced by the Australian Competition and Consumer Commission (ACCC).

According to FSANZ, labels should include the following types of information:

NUTRITION INFORMATION		
Servings per can: 2		
Serving size: 210g		
	Average Quantity Per serving	Average Quantity Per 100g
ENERGY	895kJ	425kJ
PROTEIN	10.8g	5.1g
FAT: TOTAL	1.2g	0.6g
-SATURATED	0.2g	0.1g
CARBOHYDRATE	33.7g	16.1g
-SUGARS	15.5g	7.4g
DIETARY FIBRE	11.9g	5.7g
SODIUM	1300mg	620mg
POTASSIUM	650mg	310mg
IRON	2.7mg	1.3mg

- Nutrition information panel: this provides information on the average amount of energy (kJ), protein, total fat, saturated fat, carbohydrate, sugars, dietary fibre, sodium, potassium and iron. This information must be presented in a standard format which states the average amount per serve as well as per 100 g or 100 ml of the food.

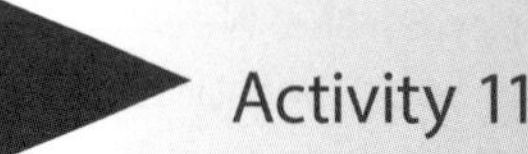

Activity 11.15: Nutrition panel

Which foods do not require a nutrition panel?

Activity 11.16: Food nutrition panels

Use the website:
www.foodstandards.gov.au/consumer/labelling/panels/Pages/default.aspx
to write notes on the following aspects of food nutrition panels:

- Serving size
- Quantity per 100g/100ml
- Energy/kJ
- Protein
- Fat
- Carbohydrates
- Sugars
- Dietary fibre
- Salt/sodium.

Alternatively, visit the FSANZ website and use their search bar to locate the information.

- Percentage labelling: this applies to most packaged foods that have to show the amount of the key or essential ingredients the product contains. For example, when comparing packaged apple pies, the apple needs to be listed with the percentage in the final product, eg. 67% apple. This allows the consumers to make comparison across the product range.

Activity 11.17: Percentage labelling

Can you think of any product that doesn't have a key or characterising ingredient in it?

- Name or description of the food: all foods must have an accurate name or description on its label that specifies the true nature of the food. For example, Chocolate Biscuits need to have chocolate in them and not any other form of flavouring that is dominant.

- Ingredients list: ingredients must be itemised on the packaging in order of weight (descending) with the main ingredient first. For example, if the chocolate biscuits had mostly chocolate in them by weight, it would be the first ingredient listed. If added water is used in the product, it must be listed most times according to its ingoing weight, with allowances made for any lost during processing like steam or evaporation. Exceptions to this rule include if the water is a minimal ingredient contributing to less than 5% of the final product weight, if it is part of another ingredient like syrup that is listed in the ingredient list or if it used to reconstitute dehydrated ingredients like apricots.

Ingredients

Contains Gluten containing ingredients, Soy and Milk as indicated in bold type.

Wholegrain Cereals (53%) [Wheat (36%), Corn (17%)], Sugar, Wheat Flour, Formulated Supplementary Food Base: MILO (10%) [Barley Malt Extract, Rice and/or Barley, Milk Solids, Sugar, Fat-Reduced Cocoa Powder, Minerals (Calcium, Magnesium and Iron), Vitamins (C, B1, B2 and A), (Soy)], Fat-Reduced Cocoa Powder, Barley Malt Extract, Skimmed Milk Powder, Sunflower Oil, Emulsifier (Soy Lecithin), Flavours (Chocolate and Vanillin), Salt.

Vitamins and Minerals

Minerals (Calcium, Iron and Zinc), Vitamins (C, Niacin, E (Soy), B1, B2, B6 and Folate).

Made on equipment that also processes products containing Oats and Tree Nuts

- Information for allergy sufferers: as mentioned in the previous chapter, food intolerances or allergies seem to be increasing. This is a concern for packaged food as some of the ingredients can cause severe allergic reactions including anaphylaxis. The common foods that cause food allergies are:
  - peanuts
  - tree nuts
  - milk
  - eggs
  - sesame seeds
  - fish and shellfish
  - soy
  - wheat.

  Each of these foods must be declared on the food label however small its quantity. Gluten is also an allergen for people with wheat intolerance or coeliac disease and its sources must be declared on product packaging.

  Another product that needs to be identified on the package is the bee product royal jelly. This has been known to cause severe allergic reactions in some people and, in rare cases, fatalities; especially in people who have asthma or suffer from other allergies.

Activity 11.18: Allergy sufferers

For someone who is intolerant to wheat or has coeliac disease, list all of the ingredients they need to avoid. What are suitable substitutes?

To protect consumers, manufacturers sometimes put on their product packaging 'may contain' for example – nuts. This is because the product may contain unintentional traces of allergens if the same equipment is used for manufacturing different products.

Food allergies are a serious concern and can be life-threatening. The only way people with a food allergy can manage it is to avoid the product entirely. This is why there are food laws that require mandatory labelling of food allergens so people can avoid those which they are allergic to.

- Date marking: otherwise known as 'use-by dates' and 'best before' dates, this type of labelling gives a guide as to how long a food product can be kept under normal conditions before it starts to deteriorate and become unsafe to eat. Any foods that are past the use-by date should not be eaten and are illegal to sell because they pose a health safety risk to consumers. The foods which have a best before date can still be eaten for a while after it has expired but may have lost some quality in terms of texture or taste. Retailers can still sell these foods as long as they are still fit for human consumption.

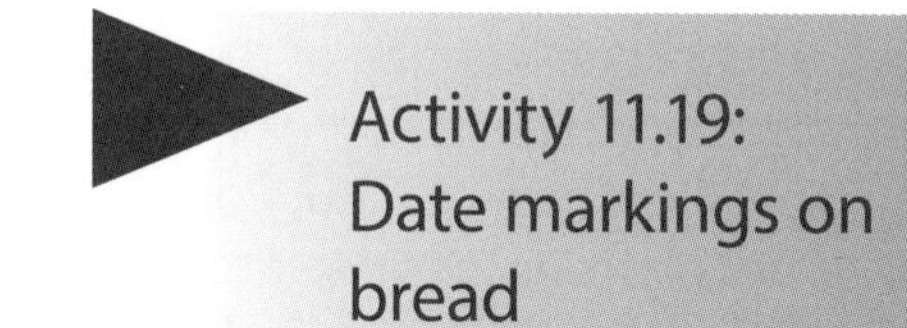

Activity 11.19: Date markings on bread

Examine the date marking for bread. How and why is this different from the examples here?

Some foods do not need to have a date marking as their shelf life extends for two years or longer as in the case of canned products. They can retain the food's quality for several years and usually are consumed before they deteriorate and become unsafe to eat.

Another requirement for suppliers when using the date marking is to detail any specific storage instructions if they are needed. Examples include dairy products that must be refrigerated or frozen goods that must be kept at specific temperatures. Any detailed cooking instructions should also be provided by the manufacturer and followed by the consumer.

Activity 11.20: Keeping times and storage conditions

Find out the average keeping times (use-by/best before dates) and storage conditions for the following products and record them in a chart such as the one below.

Product	Keeping time	Storage conditions
Bread		
Milk		
Frozen fish		

Product	Keeping time	Storage conditions
Butter		
Margarine		
Cheddar cheese		
Cream cheese		
Packaged biscuits		
Chicken pieces		
Dry breadcrumbs		
Raw cashews		
Ice cream		

- Country of origin: the Food Standards Code requires all packaged food and almost all unpackaged food sold in Australia to have a country of origin label on it somewhere. The Australian Competition and Consumer Commission (ACCC) regulates laws pertaining to country of origin labelling. Packaged food must contain a declaration identifying:
  - the country where the food was made, produced or grown; or
  - the country where the food was manufactured or produced/packaged and whether the food is a mix of local and imported ingredients or entirely made of imported ingredients. If this is the case, the country of origin of the imported ingredients can be included.

Unpackaged food needs to be labelled with the country of origin where it is sold. This applies to fish and shellfish, processed and fresh fruit, herbs, legumes, nuts, seeds, spices, vegetables, and meat including beef, chicken, pork and sheep. This can be done by either a label on the pricing tag or stickers on the product as in the case of fruit.

Activity 11.21: Country of origin

The ACCC outlines what labels a manufacturer is allowed to put on their product in reference to the country of origin. Research the following labels and describe the differences between them.

- 'Made in'
- 'Product of'
- 'Grown in'

Activity 11.22: Imported?

Examine a range of packaged products and find out their country of origin. Why do you think they were imported? Explain your response.

Activity 11.23: Symbols

What are some of the symbols that are used to identify where the product was made?

Did you know ...

Some foods include claims on their labels such as 'Proudly Australian owned' or '100% Australian owned'. These statements are about the ownership of the company; they don't indicate where the product was made or where its ingredients came from.

- Barcode: a barcode is a series of lines and spaces representing a number that is usually 8 to 13 digits in length and that is readable by a barcode scanner. They are used to identify the millions of products in the world in a very quick manner by simply scanning it to retrieve information from a software programme. The barcode is only a link between the product and the product information existing within the software. For example, the MYOB Retail Manager program can retrieve information like product description, cost price and selling price from the barcode on the product item. It is a useful tool for inventory control as retailers do not have to do a manual shelf count of products because the computer system will have that information.

  In some countries, barcodes are a legal requirement and many retailers will insist on it as part of the conditions of product supply. There are two main types of barcodes used – the 12 digit Universal Product Code (UPC) used in America and Canada, and the 13 digit European Article Numbering (EAN). Australia accepts both forms of these barcodes.

Activity 11.24: Bar code

Does the bar code have to be a specific size?

- Weights and measures: any food sold must contain either a weight or form of measure. In Australia, these are regulated by the Australian National Measurement Institute, which ensures manufacturers are being accurate about their product claims. They ensure that the packaged products:
  - have standardised metric units of measurements (ml, dL, L, mg, g, Kg or an actual number of items in the package)

- are situated on the main display site of the packaging so the buyer can easily see it
- are of a certain minimum size for legibility and clear from all other graphics or fonts
- are written in the same direction as the brand name of the product
- are not be hidden by the addition of extra packaging labels.

Any manufacturing or packaging site must have certified and standardised measuring instruments to ensure accuracy and compliance with weights and measures. The instruments must be maintained and checked regularly to ensure they are working properly.

Activity 11.25: Weights and measures

1. Examine several unopened packaged products and note down their listed measurement.
2. Weigh or decant the product into a measuring jug/cylinder and note any differences.
3. Then weigh the packaging.
4. Do differences still exist? Why do you think this is so?

- Food recall information: it is the responsibility of FSANZ to coordinate and monitor food recalls in Australia. FSANZ works with both state and territory governments and either the food product's manufacturer or importer to ensure the food recall process works quickly and efficiently. Food is recalled when there is a safety or health risk from distribution, sales and consumption. In other words, it is recalled from warehouses and shops, and consumers are asked not to use it and return the item to the store for a full refund. Food is recalled when there is a complaint – this can come from consumers, retailers, the manufacturers or from government bodies.

  If a food is recalled, it is the responsibility of the food business to work quickly to remove the product from the shelves and inform consumers. FSANZ assists in the recall process by informing the industry as well as government agencies and departments. Food Standards also assists to inform consumers via a webpage devoted to product recalls. Consumers can also follow food recalls on Twitter and Facebook, or by subscribing to Food Recall Alerts, a useful tool for allergy sufferers.

  Food Standards have surveyed the food recalled since 1990 and have come up with a classification list of the reason why they are recalled. Some of the headings include:

  - biotoxin which is a toxin formed from a living organism including plant, animals or fungi
  - chemical or contaminant present
  - incorrect labelling
  - microbial contamination
  - packaging fault

- presence of foreign matter
- tampering of the package
- undeclared allergen.

They found that between 2002 and the end of 2011, there were 663 food recalls of which 229 (35%) were caused by microbial contamination and 218 (33%) were due to incorrect labelling, which included undeclared allergens.

Activity 11.26: Research

Research and find out the difference between a 'food recall' and a 'food withdrawal'.

Activity 11.27: Food recalls

Often food recalls are advertised in newspapers. Check The West Australian newspaper over a week and see if any food has been recalled. What was it and why was it recalled? Check your responses with other class members.

- Other labelling information can include:
  - suggestions for serving either written or in the form of pictures.
  - special logos to promote health. These include the Heart Foundation's 'Health Tick'. Can you think of more?
  - the name and business address of supplier.

Activity 11.28: Food labelling

Examine four different food products and check their labels as to whether they have the types of information as mentioned previously.

Useful Websites

www.safeworkaustralia.gov.au/sites/swa/whs-information/hazardous-chemicals/sds/pages/sds: Provides information about Safety Data Sheets

www.safeworkaustralia.gov.au/sites/swa/model-whs-laws/pages/model-whs-laws: Outlines the new WHS model and laws regarding workplace safety

www.commerce.wa.gov.au/Worksafe: Contains advice about work safe laws

www.measurement.gov.au/Pages/default.aspx: Contains information about product labelling

REFERENCES

American Dietetic Association. www.eatright.org (Viewed June 25, 2014).

Aussie Cooking. www.aussiecooking.com.au (Viewed June 25, 2014).

Aussie-Info.com. www.aussie-info.com (Viewed June 25, 2014).

Australian Association of National Advertisers. www.aana.com.au (Viewed June 25, 2014).

Australian Broadcasting Commission. Health. www.abc.net.au/health/ (Viewed June 25, 2014).

Australian Broadcasting Commission. Schools Education. www.splash.abc.net.au/home (Viewed June 25, 2014).

Australian Bush Foods Magazine. www.ausbushfoods.com (Viewed June 25, 2014).

Australian Bureau of Agriculture and Resource Economics (ABARE). www.daff.gov.au/abares (Viewed June 25, 2014).

Australian Bureau of Statistics. www.abs.gov.au (Viewed June 25, 2014).

Australian Certified Organic. www.aco.net.au (Viewed June 25, 2014).

Australian Competition and Consumer Commission. www.accc.gov.au (Viewed June 24 2014).

Australian Egg Corporation Limited. www.eggs.org.au (Viewed June 25, 2014).

Australian Food News. www.ausfoodnews.com.au (Viewed June 25, 2014).

Australian Government ComLaw Trade Practices. www.comlaw.gov.au (Viewed June 24, 2014).

Australian Institute of Health and Welfare. www.aihw.gov.au/overweight-and-obesity (Viewed June 27, 2014).

Australian Institute of Sport. www.ais.org.au /nutrition (Viewed July 5, 2014).

Australian Packaging Covenant. www.packagingcovenant.org.au (Viewed June 25, 2014).

Australian Trade Commission. www.austrade.gov.au (Viewed June 25, 2014).

Better Health Channel. Victorian Government. www.betterhealth.vic.gov.au (Viewed June 25, 2014).

Boabs in the Kimberley. www.boabsinthekimberley.com.au (Viewed June 25, 2014).

BRI Research. www.bri.com.au (Viewed June 25, 2014).

Can and Aerosol News. www.can-news.com.au (Viewed June 25, 2014).

Centre for Applied Learning Systems, Adelaide College of TAFE. (1991). Principles and methods of cookery. Adelaide, South Australia: Adelaide College of TAFE.

Coles. www.colesonline.com.au (Viewed June 25, 2014).

Commonwealth Scientific and Industrial Research Organisation. www.csiro.au (Viewed June 25, 2014).

Commonwealth Scientific and Industrial Research Organisation. www.solve.csiro.au (Viewed June 25, 2014).

Corn Refiners Association. www.corn.org (Viewed June 25, 2014).

Cox, E. M., & Francis, C. E. (1988). The book of food for health and pleasure. Melbourne, Australia: Longman Cheshire.

Dairy Australia. www.dairyaustralia.com.au (Viewed June 25, 2014).

Dairy Industry Association of Australia. www.diaa.asn.au (Viewed June 25, 2014).

Department of Agriculture and Food, Western Australia. www.agric.wa.gov.au (Viewed June 25, 2014).

Department of Agriculture, Fisheries and Forestry. www.daff.gov.au (Viewed June 25, 2014).

Department of Commerce. www.commerce.wa.gov.au/Worksafe/ (Viewed June 29, 2014)

Department of Fisheries, Government of Western Australia. www.fish.wa.gov.au (Viewed June 25, 2014).

Department of Health, Government of Western Australia. www.population.health.wa.gov.au (Viewed June 25, 2014).

Department of Health and Ageing. www.healthyactive.gov.au (Viewed June 25, 2014).

Diabetes Australia. www.diabetesaustralia.com.au (Viewed June 25, 2014).

Diabetes Institute. www.diabetes.org.au/ (Viewed June 25, 2014)

Dietitians Association of Australia. www.daa.asn.au (Viewed June 25, 2014).

Fair Trade Association. www.fta.org.au (Viewed June 25, 2014).

Fair Trade International. www.fairtrade.net (Viewed June 25, 2014)

Food Additives and Ingredients Association. www.faia.org.uk (Viewed June 25, 2014).

Food Australia. www.foodaust.com.au (Viewed June 25, 2014).

Food Cents. www.foodcentsprogram.com.au (Viewed June 25, 2014).

Food Safety Information Council. www.foodsafety.asn.au (Viewed June 25, 2014).

Food Safety and Inspection Service, United States Department of Agriculture. www.fsis.usda.gov (Viewed June 25, 2014).

Food Standards Australia New Zealand (FSANZ). www.foodstandards.gov.au (Viewed June 25, 2014).

Foodwatch. www.foodwatch.com.au (Viewed June 25, 2014).

Fresh Finesse. www.freshf.com.au (Viewed June 25, 2014).

Fresh for Kids. www.freshforkids.com.au (Viewed June 25, 2014).

George Weston Food. www.georgewestonfoods.com.au (Viewed June 25, 2014).

Glaister, R. (1993). Catering towards a career. South Melbourne, Australia: Pearson Education

Australia Pty Ltd.

Go for 2 and 5. www.gofor2and5.com.au (Viewed June 25, 2014).

Go Grains. www.grdc.com.au (Viewed June 25, 2014).

Golden Eggs. www.goldeneggs.com.au (Viewed June 25, 2014).

Good Food. www.goodfood.com.au (Viewed 25th June 2014).

Government of Western Australia Public Health. www.public.health.wa.gov.au (Viewed June 28, 2014).

Grain Corp Pty Ltd. www.graincorp.com.au (Viewed June 25, 2014).

Grain Research and Development Corporation. www.grdc.com.au (Viewed June 25, 2014).

HACCP Australia. www.haccp.com.au (Viewed June 25, 2014).

Hark, L. & Deek, D. (2005). Nutrition: The definitive Australian guide to eating for good health. Camberwell, Victoria: Dorling Kindersley.

Hayter, R. (Ed.). (1989). Foodcraft 1. The dry processes. London, UK: Macmillan.

Hayter, R. (Ed.). (1989). Foodcraft 2. The wet processes. London, UK: Macmillan.

Healthinsite. www.healthinsite.gov.au (Viewed June 25, 2014).

Heart Foundation. www.heartfoundation.com.au (Viewed June 25, 2014).

Heath, G., McKenzie, H., & Tully, L. (2006). Food solutions food and technology units 1 & 2. Melbourne, Australia: Pearson Education Australia.

Heath, G., McKenzie, H., & Tully, L. (2006). Food solutions food and technology units 3 & 4. Melbourne, Australia: Pearson Education Australia.

Home Economics Association of Australia. (2003). Nutrition the inside story. Home Economics Institute of Australia.

Ingham Enterprises. www.inghams.com.au (Viewed June 25, 2014).

In Mamas Kitchen. www.inmamaskitchen.com (Viewed June 25, 2014).

Jones, C. (2007). Bread. South Yarra, Vic: Macmillan Educational.

Jones, J. (1991). The Macquarie dictionary of cookery. Australia: Macquarie Library Pty Ltd.

KitchenwareDirect. www.kitchenwaredirect.com.au (Viewed June 25, 2014).

Kitchen Gadgets. www.kitchengadgets.com.au (Viewed June 25, 2014).

Landcare Australia. www.landcareonline.com.au (Viewed June 25, 2014).

Mason, P. (2008). How big is your food footprint? South Yarra, Vic: Macmillan Library.

Meat and Livestock Australia. www.mla.com.au (Viewed June 25, 2014).

Meat and Livestock Food Safety. www.beefandlamb.com.au/How_to/Buying_beef_and_lamb (Viewed June 25, 2014).

Melbourne Market Authority. www.marketfresh.com.au (Viewed June 25, 2014).

Meredith, S., Sullivan, C., Weihen, L., & Redman, B. (2003). Heinemann@work: Hospitality book 1. Port Melbourne, Victoria: Heinemann.

Merriman Webster. www.merriman-webster.com (Viewed June 25, 2014).

Mushrooms For Life. www.powerofmushrooms.com.au (Viewed June 25, 2014).

National Association for Sustainable Agriculture Australia. www.nasaa.com.au (Viewed June 25, 2014).

National Foods. www.natfoods.com.au (Viewed June 25, 2014).

National Health and Medical Research Council. www.nhmrc.gov.au (Viewed June 25, 2014).

National Measurement Institute www.measurement.gov.au (Viewed June 29, 2014).

New South Wales Department of Health. www.health.nsw.gov.au/pages/a2z.aspx (Viewed June 25, 2014).

New South Wales Meals on Wheels Association. www.nswmealsonwheels.org.au (Viewed June 25, 2014).

Nutrition Australia. www.nutritionaustralia.org (Viewed June 25, 2014).

Organic Food Directory. www.organicfood.co.uk (Viewed June 25, 2014).

Oxfam Australia. www.oxfam.org.au (Viewed June 25, 2014).

Parmalat Australia. www.pauls.com.au (Viewed June 25, 2014).

Peckham, G. C. & Freeland-Graves, J. H. (1979). Foundations of food preparation. New York: Macmillan Publishing Co.

People for Fair Trade. www.fairtrade.asn.au (Viewed June 25, 2014).

Perraton, G., Weston, K., Carey, D., Compton, L., & Brown, M. (2000). Food and technology 1. Queensland, Australia: John Wiley & Sons.

Perraton, G., Boddy, G., Compton, L., Weston, K., & Brown, M. (2006). Food and technology 2. Queensland, Australia: John Wiley & Sons.

Pfizer Australia, Health Report. www.healthreport.com (Viewed June 25, 2014).

Potatoes. www.freshpotatoes.com.au (Viewed June 25, 2014).

Powell, J. (1995). Food. New York: Thomson Learning.

Pulse Australia. www.pulseaus.com.au (Viewed June 25, 2014).

Queensland Meals on Wheels Services Association Inc. www.qmow.org (Viewed June 25, 2014).

Quorn. www.quorn.com (Viewed June 25, 2014).

Safe Work Australia. www.safeworkaustralia.gov.au (Viewed 28th June 2014).

Sakata. www.sakata.com.au (Viewed June 25, 2014).

Seafood Western Australia. www.seafoodwesternaustralia.net (Viewed June 25, 2014).

South Australian Food Centre. www.foodwine.sa.gov.au (Viewed June 25, 2014).

Sports Dieticians Australia. www.sportsdietitians.com.au (Viewed June 25, 2014).

Sullivan, C., Redman, B., & Meredith, S. (2003). Heinemann@work: Hospitality book 2. Port Melbourne, Victoria: Heinemann.

Taste.com.au, News Magazines. www.taste.com.au (Viewed June 25, 2014).

The Organic Food & Produce Company. www.organicfood.com.au (Viewed June 25, 2014).

Weihen, L., Aduckiewicz, J., & Amys, J. (1995). Investigating food technology. Port Melbourne, Victoria: Heinemann.

World Vision Australia. www.worldvision.com.au (Viewed June 25, 2014).